SCIENTIFIC GENIUS
AND CREATIVITY

Readings from
**SCIENTIFIC
AMERICAN**

SCIENTIFIC GENIUS
AND CREATIVITY

With an introduction by
Owen Gingerich
*Harvard University and
Smithsonian Astrophysical Observatory*

W. H. Freeman and Company
New York

Library of Congress Cataloging-in-Publication Data

Scientific genius and creativity.

Bibliography: p.
Includes index.
1. Scientists—Biography. 2. Creative ability in science. I. Gingerich, Owen. II. Scientific American.
Q141.S293 1987 509.2′2 [B] 86-31920

ISBN 0-7167-1858-8

Printed in the United States of America

9 8 7 6 5 4 3 2 1 KP 5 4 3 2 1 0 8 9 8 7

CONTENTS

INTRODUCTION

This collection of biographies, selected from over sixty published in SCIENTIFIC AMERICAN over the past few decades, speaks for itself. It tells colorful stories of a few of the physicists, chemists, doctors, biologists and geologists who have built our scientific edifice since the Renaissance. Indeed, these essays celebrate the short biography as an art form.

Together these articles raise a provocative question about the role of scientific genius in the shaping of science. As we reflect on the pioneers of the past, we begin to appreciate that they truly were "makers of Western thought" and their perceptions of nature strongly condition our own views. Looking into the future, we can expect a new science to emerge, because if the past is any guide, we can be assured that the unique insights of new pioneers will give us fresh ways of seeing old knowledge and new discoveries will reveal previously unperceived phenomena.

I remember Cecilia Payne-Gaposchkin telling me that in her autobiography she had tried to suppress her own formative role in astrophysics because "if I hadn't found it, some one else would have." I only half believe that, and I think her book is poorer for her all-too-humble attitude. What she did and what many other scientific geniuses have done came at unique times and were presented in unique ways. Our science has been molded, perhaps as much by this historical progression of personal achievement and understanding, as by any intrinsic patterns of nature waiting to be perceived and analyzed.

At the beginning and conclusion of this collection I have deliberately placed two articles that probe the nature of scientific creativity and discovery. These essays frame the biographies and concentrate our philosophical attention on the stories that comprise the main substance of this Reader. Both Jacob Bronowski in "The Creative Process" and Gunther Stent in "Prematurity and Uniqueness in Scientific Discovery" remark on the creative activity that forms our scientific world view and compare these scientific accomplishments with the greatest achievements of such artistic geniuses as Shakespeare and Beethoven. Without Shakespeare there would be no *King Lear*, without Beethoven no *Fifth Symphony*. Would we have Newton's laws without Newton? Perhaps, but probably quite differently formulated. (Physicists will remember that Newton's first law, for example, has been criticized as redundant.) And certainly there would be no *Principia* without Newton. The arguments by Bronowski and Stent seem to me well made, and I invite you, the reader, to consider the nature of scientific creativity as you examine these individual accounts.

These biographies have been chosen for their diversity and readability, and many of them bring insights from leading historians of science. Whether or not you are prepared to ponder the impact that an individual scientist has on the growth of science, I hope you will find these articles both fascinating and rewarding.

Owen Gingerich
Harvard-Smithsonian Center
for Astrophysics

November 1986

SCIENTIFIC GENIUS
AND CREATIVITY

The Creative Process

by J. Bronowski
September 1958

Introducing an issue on innovation in science. The argument: Although science and art are social phenomena, an innovation in either field occurs only when a single mind perceives in disorder a deep new unity

The most remarkable discovery made by scientists is science itself. The discovery must be compared in importance with the invention of cave-painting and of writing. Like these earlier human creations, science is an attempt to control our surroundings by entering into them and understanding them from inside. And like them, science has surely made a critical step in human development which cannot be reversed. We cannot conceive a future society without science.

I have used three words to describe these far-reaching changes: discovery, invention and creation. There are contexts in which one of these words is more appropriate than the others. Christopher Columbus discovered the West Indies, and Alexander Graham Bell invented the telephone. We do not call their achievements creations because they are not personal enough. The West Indies were there all the time; as for the telephone, we feel that Bell's ingenious thought was somehow not fundamental. The groundwork was there, and if not Bell then someone else would have stumbled on the telephone as casually as on the West Indies.

By contrast, we feel that *Othello* is genuinely a creation. This is not because *Othello* came out of a clear sky; it did not. There were Elizabethan dramatists

CAVE PAINTING is a profound innovation which can be compared with the invention of writing and of science. The paintings on the opposite page adorn the roof and the walls of a cave in Montignac in south-central France. The horned bull at upper left and the troop of horses beneath it were painted at least 15,000 years ago. The animals toward the rear of the roof were painted at least 17,000 years ago. This photograph, made by *Life* photographer Ralph Morse, is published by the courtesy of *Life*.

before Shakespeare, and without them he could not have written as he did. Yet within their tradition *Othello* remains profoundly personal; and though every element in the play has been a theme of other poets, we know that the amalgam of these elements is Shakespeare's; we feel the presence of his single mind. The Elizabethan drama would have gone on without Shakespeare, but no one else would have written *Othello*.

There are discoveries in science like Columbus's, of something which was always there: the discovery of sex in plants, for example. There are tidy inventions like Bell's, which combine a set of known principles: the use of a beam of electrons as a microscope, for example. In this article I ask the question: Is there anything more? Does a scientific theory, however deep, ever reach the roundness, the expression of a whole personality that we get from *Othello*?

A fact is discovered, a theory is invented; is any theory ever deep enough for it to be truly called a creation? Most nonscientists would answer: No! Science, they would say, engages only part of the mind—the rational intellect—but creation must engage the whole mind. Science demands none of that groundswell of emotion, none of that rich bottom of personality, which fills out the work of art.

This picture by the nonscientist of how a scientist works is of course mistaken. A gifted man cannot handle bacteria or equations without taking fire from what he does and having his emotions engaged. It may happen that his emotions are immature, but then so are the intellects of many poets. When Ella Wheeler Wilcox died, having published poems from the age of seven, *The Times* of London wrote that she was "the most

popular poet of either sex and of any age, read by thousands who never open Shakespeare." A scientist who is emotionally immature is like a poet who is intellectually backward: both produce work which appeals to others like them, but which is second-rate.

I am not discussing the second-rate, and neither am I discussing all that useful but commonplace work which fills most of our lives, whether we are chemists or architects. There are in my laboratory of the British National Coal Board about 200 industrial scientists—pleasant, intelligent, sprightly people who thoroughly earn their pay. It is ridiculous to ask whether they are creators who produce works that could be compared with *Othello*. They are men with the same ambitions as other university graduates, and their work is most like the work of a college department of Greek or of English. When the Greek departments produce a Sophocles, or the English departments produce a Shakespeare, then I shall begin to look in my laboratory for a Newton.

Literature ranges from Shakespeare to Ella Wheeler Wilcox, and science ranges from relativity to market research. A comparison must be of the best with the best. We must look for what is created in the deep scientific theories: in Copernicus and Darwin, in Thomas Young's theory of light and in William Rowan Hamilton's equations, in the pioneering concepts of Freud, of Bohr and of Pavlov.

The most remarkable discovery made by scientists, I have said, is science itself. It is therefore worth considering the history of this discovery, which was not made all at once but in two periods. The first period falls in the great age of Greece, between 600 B.C. and 300 B.C.

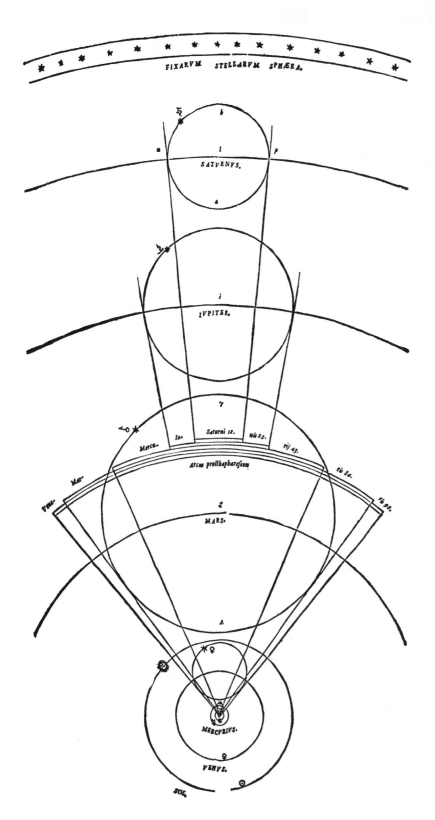

PRE-COPERNICAN CONCEPTION of the solar system is depicted in this woodcut from Copernicus's successor Johannes Kepler. The earth is in the center (*bottom*). Around it move Mercury, Venus, the sun (*Sol*), Mars, Jupiter and Saturn. At top is the sphere of the fixed stars (*fixarum stellarum sphaera*). The planets moved not only in orbits but also in epicycles centered on the orbits. The straight lines show angles subtended by epicycles.

The second period begins roughly with the Renaissance, and is given impetus at several points by the rediscovery of Greek mathematics and philosophy.

When one looks at these two periods of history, it leaps to the eye that they were not specifically scientific. On the contrary: Greece between Pythagoras and Aristotle is still, in the minds of most scholars, a shining sequence of classical texts. The Renaissance is still thought of as a rebirth of art, and only specialists are uncouth enough to link it also with what is at last being called, reluctantly, the Scientific Revolution. The accepted view of Greece and of the Renaissance is that they were the great creative periods of literature and art. Now that we recognize in them also the two periods in which science was born, we must surely ask whether this conjunction is accidental. Is it a coincidence that Phidias and the Greek dramatists lived in the time of Socrates? Is it a coincidence that Galileo shared the patronage of the Venetian republic with sculptors and painters? Is it a coincidence that, when Galileo was at the height of his intellectual power, there were published in England in the span of 12 years the following three works: the Authorized Version of the Bible, the First Folio of Shakespeare and the first table of logarithms?

The sciences and the arts have flourished together. And they have been fixed together as sharply in place as in time. In some way both spring from one civilization: the civilization of the Mediterranean, which expresses itself in action. There are civilizations which have a different outlook; they express themselves in contemplation, and in them neither science nor the arts are practiced as such. For a civilization which expresses itself in contemplation values no creative activity. What it values is a mystic immersion in nature, the union with what already exists.

The contemplative civilization we know best is that of the Middle Ages. It has left its own monuments, from the Bayeux Tapestry to the cathedrals; and characteristically they are anonymous. The Middle Ages did not value the cathedrals, but only the act of worship which they served. It seems to me that the works of Asia Minor and of India (if I understand them) have the same anonymous quality of contemplation, and like the cathedrals were made by craftsmen rather than by artists. For the artist as a creator is personal; he cannot drop his work and have it taken up by another without doing it violence. It may

NICOLAI COPERNICI

net, in quo terram cum orbe lunari tanquam epicyclo contineri diximus. Quinto loco Venus nono mense reducitur. Sextum deniqʒ locum Mercurius tenet, octuaginta dierum spacio circū currens. In medio uero omnium residet Sol. Quis enim in hoc

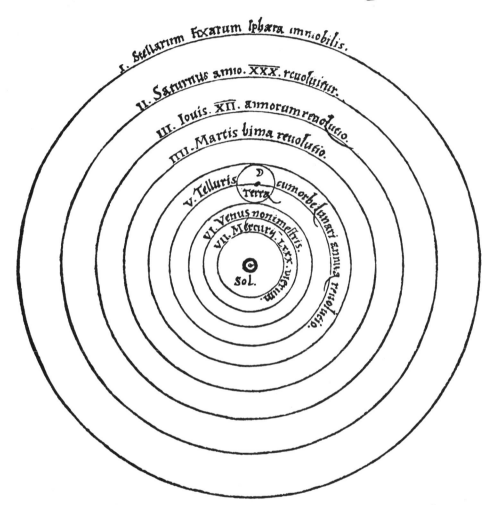

pulcherimo templo lampadem hanc in alio uel meliori loco po neret, quàm unde totum simul possit illuminare? Siquidem non inepte quidam lucernam mundi, alij mentem, alij rectorem uo= cant. Trimegistus uisibilem Deum, Sophoclis Electra intuentē omnia. Ita profecto tanquam in solio re gali Sol residens circum agentem gubernat Astrorum familiam. Tellus quoqʒ minime fraudatur lunari ministerio, sed ut Aristoteles de animalibus ait, maximā Luna cū terra cognationē habet. Concipit interea à Sole terra, & impregnatur annuo partu. Inuenimus igitur sub
hac

COPERNICAN CONCEPTION of the solar system is repro- duced from Copernicus's own *De Revolutionibus Orbium* *Coelestium*. The sun is in the center. Around it move Mercury, Venus, the earth (Telluris) and moon, Mars, Jupiter and Saturn.

be odd to claim the same personal engagement for the scientist; yet in this the scientist stands to the technician much as the artist stands to the craftsman. It is at least remarkable that science has not flourished either in an anonymous age, such as the age of medieval crafts, or in an anonymous place, such as the craftsmanlike countries of the East.

The change from an outlook of contemplation to one of action is striking in the long transition of the Renaissance and the Scientific Revolution. The new men, even when they are churchmen, have ideals which are flatly opposed to the monastic and withdrawn ideals of the Middle Ages. Their outlook is active, whether they are artists, humanist scholars or scientists.

The new man is represented by Leonardo da Vinci, whose achievement has never, I think, been rightly understood. There is an obvious difference between Leonardo's painting and that of his elders—between, for example, an angel painted by him and one by Verrocchio. It is usual to say that Leonardo's angel is more human and more tender; and this is true, but it misses the point. Leonardo's pictures of children and of women are human and tender; yet the evidence is powerful that Leonardo liked neither children nor women. Why then did he paint them as if he were entering their lives? Not because he saw them as people, but because he saw them as expressive parts of nature. We do not understand the luminous and transparent affection with which Leonardo lingers on a head or a hand until we look at the equal affection with which he paints the grass and the flowers in the same picture.

To call Leonardo either a human or a naturalist painter does not go to the root of his mind. He is a painter to whom the detail of nature speaks aloud; for him, nature expresses herself in the detail. This is a view which other Renaissance artists had; they lavished care on perspective and on flesh tones because these seemed to them (as they had not seemed in the Bayeux Tapestry) to carry the message of nature. But Leonardo went further; he took this artist's vision into science. He understood that science as much as painting has to find the design of nature in her detail.

When Leonardo was born in 1452, science was still Aristotle's structure of cosmic theories, and the criticism of Aristotle in Paris and Padua was equally grandiose. Leonardo distrusted all large theories, and this is one reason why his

experiments and machines have been forgotten. Yet he gave science what it most needed, the artist's sense that the detail of nature is significant. Until science had this sense, no one could care—or could think that it mattered—how fast two unequal masses fall and whether the orbits of the planets are accurately circles or ellipses.

The power which the scientific method has developed has grown from a procedure which the Greeks did not discover: the procedure of induction. This procedure is useless unless it is followed into the detail of nature; its discovery therefore flows from Leonardo's vision.

Francis Bacon in 1620 and Christian Huygens in 1690 set down the intellectual bases of induction. They saw that it is not possible to reach an explanation of what happens in nature by deductive steps. Every explanation goes beyond our experience and thereby becomes a speculation. Huygens says, and philosophers have sheepishly followed him in this, that an explanation should therefore be called probable. He means that no induction is unique; there is always a set—an infinite set—of alternatives between which we must choose.

The man who proposes a theory makes a choice—an imaginative choice which outstrips the facts. The creative activity of science lies here, in the process of induction. For induction imagines more than there is ground for and creates relations which at bottom can never be verified. Every induction is a speculation and it guesses at a unity which the facts present but do not strictly imply.

To put the matter more formally: A scientific theory cannot be constructed from the facts by any procedure which can be laid down in advance, as if for a machine. To the man who makes the theory, it may seem as inevitable as the ending of *Othello* must have seemed to Shakespeare. But the theory is inevitable only to him; it is his choice, as a mind and as a person, among the alternatives which are open to everyone.

There are scientists who deny what I have said—that we are free to choose between alternative theories. They grant that there are alternative theories, but they hold that the choice between them is made mechanically. The principle of choice, in their view, is Occam's Razor: we choose, among the theories which fit the facts we know now, that one which is simplest. On this view, Newton's laws were the simplest theory which covered the facts of gravitation as they were then

known; and general relativity is not a new conception but is the simplest theory which fits the additional facts.

This would be a plausible view if it had a meaning. Alas, it turns out to be a verbal deception, for we cannot define simplicity; we cannot even say what we mean by the simpler of two inductions. The tests which have been proposed are hopelessly artificial and, for example, can compare theories only if they can be expressed in differential equations of the same kind. Simplicity itself turns out to be a principle of choice which cannot be mechanized.

Of course every innovator has thought that his way of arranging the facts is particularly simple, but this is a delusion. Copernicus's theory in his day was not simple to others, because it demanded two rotations of the earth—a daily one and a yearly one—in place of one rotation of the sun. What made his

EIGHT MIGHTY CREATORS mentioned in this article are depicted in this drawing

theory seem simple to Copernicus was something else: an esthetic sense of unity. The motion of all the planets around the sun was both simple and beautiful to him, because it expressed the unity of God's design. The same thought has moved scientists ever since: that nature has a unity, and that this unity makes her laws seem beautiful in simplicity.

The scientist's demand that nature shall be lawful is a demand for unity. When he frames a new law, he links and organizes phenomena which were thought different in kind; for example, general relativity links light with gravitation. In such a law we feel that the disorder of nature has been made to reveal a pattern, and that under the colored chaos there rules a more profound unity.

A man becomes creative, whether he is an artist or a scientist, when he finds a new unity in the variety of nature. He does so by finding a likeness between things which were not thought alike before, and this gives him a sense both of richness and of understanding. The creative mind is a mind that looks for unexpected likenesses. This is not a mechanical procedure, and I believe that it engages the whole personality in science as in the arts. Certainly I cannot separate the abounding mind of Thomas Young (which all but read the Rosetta Stone) from his recovery of the wave theory of light, or the awkwardness of J. J. Thomson in experiment from his discovery of the electron. To me, William Rowan Hamilton drinking himself to death is as much part of his prodigal work as is any drunken young poet; and the childlike vision of Einstein has a poet's innocence.

When Max Planck proposed that the radiation of heat is discontinuous, he seems to us now to have been driven by nothing but the facts of experiment. But we are deceived; the facts did not go so far as this. The facts showed that the radiation is not continuous; they did not show that the only alternative is Planck's hail of quanta. This is an analogy which imagination and history brought into Planck's mind. So the later conflict in quantum physics between the behavior of matter as a wave and as a particle is a conflict between analogies, between poetic metaphors; and each metaphor enriches our understanding of the world without completing it.

In *Auguries of Innocence* William Blake wrote:

A dog starv'd at his Master's gate
Predicts the ruin of the State.

This seems to me to have the same imaginative incisiveness, the same un-

by Eric Mose. At left is Leonardo da Vinci. Second from left is William Blake; third, William Shakespeare; fourth, Nikolaus Copernicus; fifth, Galileo Galilei; sixth, Christian Huygens. At the upper right are Albert Einstein (*left*) and Max Planck (*right*).

BAYEUX TAPESTRY, a short section of which is shown here, exemplifies the art of the "contemplative civilizations" discussed by Bronowski. Made by anonymous French artists around 1077, the tapestry is 20 inches high and 231 feet long. It depicts scenes in the life of Harold II. Here, Harold is told of an Omen (Halley's Comet). *(The Bettmann Archive)*

derstanding crowded into metaphor, that Planck had. And the imagery is as factual, as exact in observation, as that on which Planck built; the poetry would be meaningless if Blake used the words "dog," "master" and "state" less robustly than he does. Why does Blake say dog and not cat? Why does he say master and not mistress? Because the picture he is creating depends on our factual grasp of the relation between dog and master. Blake is saying that when the master's conscience no longer urges him to respect his dog, the whole society is in decay. This profound thought came to Blake again and again: that a morality expresses itself in what he called its Minute Particulars—that the moral detail is significant of a society. As for the emotional power of the couplet, it comes, I think, from the change of scale between the metaphor and its application: between the dog at the gate and the ruined state. This is why Blake, in writing it, seems to me to transmit the same excitement that Planck felt when he discovered, no, when he created, the quantum.

One of the values which science has made natural to us is originality; as I said earlier, in spite of appearances science is not anonymous. The growing

tradition of science has now influenced the appreciation of works of art, so that we expect both to be original in the same way. We expect artists as well as scientists to be forward-looking, to fly in the face of what is established, and to create not what is acceptable but what will become accepted. One result of this prizing of originality is that the artist now shares the unpopularity of the scientist: the large public dislikes and fears the way that both of them look at the world.

As a more important result, the way in which the artist looks at the world has come close to the scientist's. For example, in what I have written science is pictured as preoccupied less with facts than with relations, less with numbers than with arrangement. This new vision, the search for structure, is also marked in modern art. Abstract sculpture often looks like an exercise in topology, exactly because the sculptor shares the vision of the topologist.

I underline this common vision because I believe that history will look back on it as characteristic of our age. A hundred years ago the way to advance physics and chemistry seemed to be by making more and more exact measurements. Science then was a quantitative affair, and this 19th-century picture of the scientist preoccupied with numbers is still

large in the popular mind. But in fact the concern of science in our age is different; it is with relation, with structure and with shape.

In these articles I find my view, that a theory is the creation of unity in what is diverse by the discovery of unexpected likenesses. Innovation is pictured as an act of imagination, a seeing of what others do not see.

There is, however, one striking division, between the articles which treat physical and those which treat the biological sciences. The physical scientists have more fun. Their theories are more eccentric; they live in a world in which the unexpected is everyday. This is a strange inversion of the way that we usually picture the dead and the living, and it reflects the age of these sciences. The physical sciences are old, and in that time the distance between fact and explanation has lengthened; their very concepts are unrealistic. The biological sciences are young, so that fact and theory look alike; the new entities which have been created to underlie the facts are still representational rather than abstract. One of the pleasant thoughts that these articles prompt is: How much more extravagant the biological sciences will become when they are as old as the physical sciences.

"TWO FORMS," by the British sculptor Henry Moore, is an example of experimental modern art, the attitude of which Bronowski compares with that of modern science. This sculpture, carved in pynkado wood, is in the collection of the Museum of Modern Art.

Corn. Jansan pinx.

J. Hall sculp. London.

Guilielmus Harveius,

COLLEG. MEDICOR. LONDIN. SOCIUS.

Epictura Archetypa in Ædibus Collegii Medicorum

Londinensis asservata.

2

William Harvey

by Frederick G. Kilgour
June 1952

The 17th-century English physician who discovered the circulation of the blood was a poor experimenter but laid the basis of modern biology and medicine

 ND I remember that when I asked our famous Harvey, *in the only Discourse I had with him,* (which was but a while be fore he dyed). What were the things that induc'd him to think of a Circulation of the Blood? He answer'd me, that when he took notice that the Valves in the Veins of so many several Parts of the Body, were so Plac'd that they gave free passage to the Blood Towards the Heart, but oppos'd the passage of the Venal Blood the Contrary way: He was invited to imagine, that so Provident a Cause as Nature had not so Plac'd so many Valves without design: and no Design seem'd more probable than that, since the Blood could not well, because of the interposing Valves, be sent by the Veins to the Limbs; it should be sent through the Arteries, and Return through the Veins, whose Valves did not oppose its course that way.

The Irish chemist Robert Boyle reported this interview with William Harvey in his *Disquisition about Final Causes of Natural Things,* published 31 years after Harvey's death. It is the only recorded statement from Harvey of the clue that led him to his great discovery—an Alpine peak in the history of biology. This man who laid the basis of modern medicine is hardly known today except as a name. His classic work, written in Latin and titled *Exercitatio Anatomica de Motu Cordis et Sanguinis in Animalibus* (Anatomical Studies on the Motion of the Heart and Blood in Animals), is widely celebrated but little read. Both the man and the work are actually much more interesting than their conspicuous obscurity might suggest.

"Our famous *Harvey*" was born of yeoman stock at Folkestone in 1578; his

PORTRAIT OF HARVEY appears in an edition of his works published in 1766, a century after he had died.

father later became mayor of the city. Harvey was a schoolboy of 10 when the Spanish Armada sailed against England; he set up practice as a physician in London in the last year of Elizabeth's reign; he gave his first lecture on the circulation of the blood in 1616, the year that Shakespeare died. Like Shakespeare, Harvey left us his works but not very much about himself. Most of our knowledge about his person and character derives from the librarian and biographer John Aubrey, who wrote a "Brief Life" on him. Harvey, says Aubrey, was a very short man with a "little eie, round, very black, full of spirit." He was temperamental and had his eccentricities. As a young blood he wore a dagger in the fashion of the day and was wont to draw it on slight provocation. It is a matter of record that he married at the age of 26, but nothing is known about his wife or family life, except that he had no children. In his later years (he lived to the age of 79) Harvey liked to be in the dark, because he could think better, and he had underground caves constructed at his house in Surrey for meditation.

Harvey is known to have been a copious scribbler. He wrote hastily, and all but illegibly, in a mixture of Latin and English, and was a careless speller: in one place in his notes appears the word "piggg"—an unusually liberal number of "g's" even for 17th-century English. Aside from his classic *De Motu Cordis,* few of his writings survive. One reason is that he lost many of his papers during the Civil War of 1642, when rioters looted his house in London and destroyed his manuscripts while he was away in Nottingham with Charles I, to whom Harvey was Physician-in-Ordinary. Harvey later said this loss was the most crucifying he had ever experienced.

A dynamic little man, he spent his life in the ardent pursuit of learning, and he wrote at least a dozen treatises on various subjects, but these, like the manuscripts destroyed by the looters,

were never published and none is now known to exist. Of his few published works perhaps the most important, next to *De Motu Cordis,* was *De Generatione* (On Reproduction), which made several valuable contributions to embryology.

HIS WORK on the circulation of the blood remains, however, his one great monument. It was remarkable not only as a history-making discovery but as a pioneering expression of the scientific method in biology. Harvey was a contemporary of Galileo, Kepler, Bacon and Descartes. The scientific revolution of the Renaissance, which swept away the systems of classical philosophy and established the methods upon which modern science is based, found in him one of its earliest prodigies. Harvey was the first biologist to use quantitative methods to demonstrate an important discovery. To weigh, to measure, to count and thus to arrive at truth was such a new idea in the 17th century that even a man of Harvey's genius could do it only badly. But his application of quantitative procedures to biology ushered in the modern age for that science.

Harvey graduated from Cambridge University in 1597 and went to study medicine at the University of Padua, the greatest scientific school of the day. The anatomy and physiology of the heart, arteries, veins and blood then being taught was still mainly the system that had been constructed 14 centuries earlier by the Greek physician Galen. According to Galen, chyle (a kind of lymph) passed from the intestines to the liver, which converted it to venous blood and at the same time added a "natural spirit." The liver then distributed this blood through the venous system, including the right ventricle of the heart. Galen knew from experiment that when he severed a large vein or artery in an animal, blood would drain off from both the veins and the arteries. He realized, therefore, that there must be some connection between the veins and the arter-

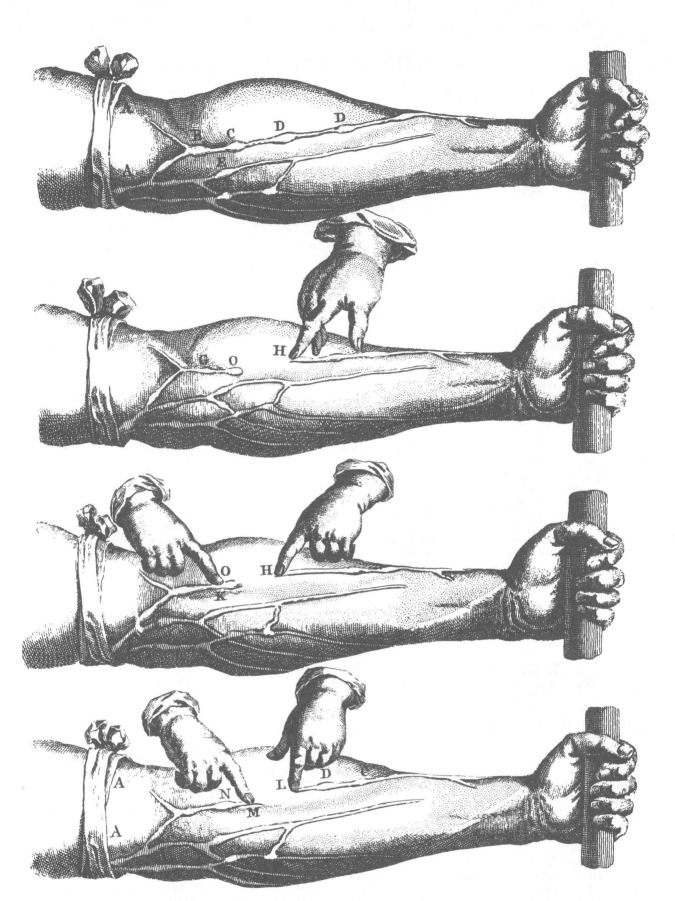

EXPERIMENTS ON THE ARM are depicted in engravings made from those in Harvey's original book for the edition of 1766. The top drawing shows that if the arm is tied off above the elbow (A), nodes appear at the location of valves in the veins (BCDEF). The second drawing shows that if the blood is pressed out of the vein between O and H and a finger placed at H, the vein remains empty. The third drawing shows that if a second finger is then pushed down the vein toward O, the vein still remains empty. The fourth drawing shows that if a finger is placed at L and a second finger pushed along the vein from L to N, the vein will likewise remain empty. By such experiments Harvey deduced that the blood flowed along the veins toward the heart.

ies, and he believed he had found such a connection in the form of pores in the wall dividing the left side of the heart from the right. He argued that the venous blood oozed through these supposed pores to the left heart, was there charged with *vital spirit* coming from the lungs and thus took on the bright crimson color of arterial blood. According to the Galenic scheme, blood flowed to various parts of the body through both the veins and the arteries to supply the body members with nourishment and spirit. There was no real circulation or motive power; the blood in the vessels simply ebbed back occasionally to the heart and lungs for the removal of impurities.

To Galen's scheme two important modifications had been made by Harvey's time. Andreas Vesalius of Padua, the founder of modern anatomy, had announced in 1555 that Galen's "pores" did not exist, and Vesalius' successor Realdo Colombo had discovered the system whereby the blood flows from the right side of the heart through the pulmonary arteries to the lungs and thence *via* the pulmonary veins to the left side of the heart. He showed by animal experiments that the pulmonary veins contain arterial blood, not "vital spirit." The second important discovery, made by Fabricius ab Aquapendente at Padua, had been that the veins possess valves—"little doors," he called them. Fabricius did not realize their function; he suggested, following Galen's ideas, that they were designed to slow the flow of blood into the extremities.

H ARVEY, with his doctor's degree from Padua, returned to England in 1602. Whether or not he had begun to form his notion of the circulation of the blood when he left Padua we do not know. In any event, he proceeded to practice medicine in London and rose rapidly in his profession. In 1615 the Royal College of Physicians, of which he was a fellow, honored him with the lifetime post of Lumleian Lecturer. In his first series of Lumleian lectures, given in 1616, he began to describe the circulation of the blood. We have his 98-page set of notes outlining these lectures. In them he describes some of his experiments, including the one which satisfied him "that so Provident a Cause as Nature had not Plac'd so many Valves without design" and which gave him the idea of the circulation, as he later told Robert Boyle.

The notes make clear that Harvey was already convinced that the blood circulates through the body and that the heart is its pumping engine. He concluded his 1616 series of lectures with this statement:

"It is proved by the structure of the heart that the blood is continuously transferred through the lungs into the aorta, as by two clacks of a water bellows to raise water. It is proved by the ligature that there is a passage of blood from the arteries to the veins. It is therefore demonstrated that the continuous movement of the blood in a circle is brought about by the beat of the heart. Is this for the sake of nutrition, or the better preservation of the blood and members by the infusion of heat, the blood in turn being cooled by heating the members and heated by the heart?"

Twelve years later Harvey, having carried out further experiments to prove his circulation theory, published *De Motu Cordis*. It is a book of only 72 pages. The volume contains two dedications (to King Charles and to Doctor Argent, President of the Royal College), an introduction and 17 brief chapters presenting his arguments.

After giving in Chapter I his reasons for writing the book (among them the desire to protect himself from ridicule), Harvey devoted the next four chapters to a remarkable analysis of the movements of the heart, arteries and auricles and an equally remarkable analysis of the functions of the heart. He had despaired at first of ever understanding the movement of the heart in warm-blooded animals, because its pulsation was so rapid, but he had found that he could analyze heart motions in cold-blooded animals and in dying warm-blooded ones. So far as direct inspection is concerned, such observations are still our principal sources for knowledge of heart motion.

Harvey gave the first clear statement of the apex beat, of the muscular character of the heart, and of the origin of the heartbeat in the right auricle and its conduction to the other auricle and the ventricles. He also demonstrated that the pulse in the arteries is due to the impact of blood ejected by the heart, as when "one blows into a glove," an image he used first in his 1616 lectures. Harvey correctly concluded that "the principal function of the heart is the transmission and pumping of the blood through the arteries to the extremities of the body."

He went on to review the movement of the blood from the right side of the heart through the lungs to the left side of the heart, as Colombo had described it, and to demonstrate how the blood passes from the left heart through the arteries to the extremities and thence *via* the veins back to the right heart. This section of the book contains the core of Harvey's discovery. He employed

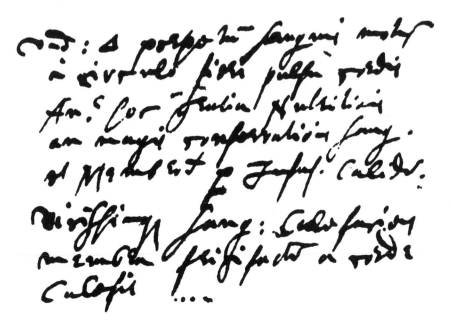

DISCOVERY OF THE CIRCULATION of the blood was first recorded by Harvey in this illegible handwriting in one of his many notebooks.

ANATOMICAL
EXERCISES,

CONCERNING
The motion of the *Heart,*
and *Blood,* in Living Creatures.

ENGLISH TRANSLATION of Harvey's *Exercitatio Anatomica de Motu Cordis et Sanguinis in Animalibus* was published in London in 1653.

three "propositions" to prove that the blood must circulate: 1) the amount of blood transmitted from the veins to the arteries is so great that all the blood in the body must pass through the heart in a short time; this quantity could not be produced by the food consumed, as Galen held; 2) the amount of blood going to the extremities is much greater than needed for the nutrition of the body; 3) the blood continuously returns to the heart from the extremities through the veins.

IT WAS to prove the first proposition that Harvey engaged in his famous quantitative work—the determination of the volume of blood pumped by the heart. To make this calculation, he had to measure the amount of blood that the heart ejects with each beat and to establish the pulse rate. The measurement of cardiac output is a tough problem, and even today there are wide variations in the measurements obtained by the various methods. But Harvey got a figure that is only one-eighteenth of the lowest estimate used today. How could he have arrived at such a ridiculously incorrect figure and at the same time have used it successfully to demonstrate such an important discovery?

Harvey based his reckoning on the fact that in a cadaver he once examined the left ventricle of the heart held more than two ounces of blood. (It must have been a dilated heart.) He assumed that between contractions the ventricle might hold as little as an ounce and a half. Assuming further that the ventricle with

each contraction ejects "a fourth, a fifth, sixth or only an eighth" of its contents (we now think it ejects nearly all), he finally calculated that the cardiac output must be at least 3.9 grams per beat. According to a present-day estimate, it is actually in the neighborhood of 89 grams. Harvey can certainly be excused for not obtaining any close estimate of the output of the human heart, but he got virtually as poor results when he tried to measure the output of a sheep's heart. If he had severed a sheep's aorta and weighed the amount of blood ejected in one minute while counting the heartbeats for that minute, he could have obtained a reasonably accurate figure for cardiac output in sheep. But he never performed that obvious experiment.

Harvey also missed the mark widely in his other important measurement—the pulse rate. Somehow he counted it to be 33 per minute, about half the actual average rate, and although he obtained other values, he generally used this figure. We cannot explain this error on the ground that it is a difficult measurement to make; why he went so wrong will always be a mystery. With his two estimates—3.9 grams for cardiac output and 33 pulse beats per minute—he obtained a figure for the rate of blood flow which is one-36th of the lowest value accepted today. One of his calculations reads: "In half an hour the heart will make 1,000 beats, in some as many as two, three, and even four thousand. Multiplying the number of drams ejected by the number of beats there

will be in half an hour either 3,000 drams, 2,000 drams, 500 ounces or some proportionate quantity of blood transferred into the arteries by the heart, but always a larger quantity than is contained in the whole body." In this summation, the lowest weight, 2,000 drams, equals 17.1 pounds, which is well in excess of the total of 15 pounds of blood contained by an average human body weighing 150 pounds.

Even with his faulty calculations, Harvey proved his main point: that each half-hour the blood pumped by the heart far exceeds the total weight of blood in the body. This was a blow to the Galenic concept, for it was obvious that the food a man eats could not produce blood continuously in any such volume.

Less impressive was Harvey's demonstration of the second proposition: that the amount of blood going to the extremities is much greater than is needed for the nutrition of the body. He used no specific measurements and argued largely by inference. However, in this discussion he made the important point that the blood must pass from the arteries to the veins in the extremities. And he described here the experiment that had first suggested the circulation idea. By employing a bandage in such a way as to stop the flow in the veins while leaving the arteries open, he showed that the veins would swell but not the arteries. When he increased pressure enough to cut off the arteries as well, the veins did not swell. From these observations, Harvey reasoned correctly that the blood entered the extremities through the arteries and passed somehow to the veins. He later looked for the channels of connection but could not find them.

By another historic experiment Harvey proved his third proposition: that the blood flows in the veins toward the heart and not away from it, as the Galenic concept held. He showed that if one pressed a finger on a vein and moved the finger along the vein from below one valve to above the next, the blood thus pushed up the vein did not return to the emptied section. In short, the valves were one-way; the blood did not flow back and forth in the venous system.

WHAT ARE the principal features of Harvey's discovery? The essential factors of the cardiovascular system which effect the circulation of the blood are the pumping heart, the passage of the blood from one side of the heart through the lungs to the other side and its subsequent passage through the arteries to every part of the body and back through the veins to the heart again. Harvey already knew of the passage through the lungs when he began his work; his great contribution was to demonstrate the circulation through the arteries and veins and to integrate it with the pulmonary passage, thus establishing one comprehensive system for the movement of blood through the body. There remained, of course, one final uncharted link: How did the blood pass from the arteries to the veins in the extremities in order to return to the heart? Thirty-three years after the appearance of *De Motu Cordis* the Italian anatomist Marcello Malpighi filled in that link by discovering the capillaries and so completed Harvey's scheme.

The direct contributions of Harvey's discovery to medicine and surgery are obviously beyond measuring: it is the basis for all work in the repair of damaged or diseased blood vessels, the surgical treatment of high blood pressure and coronary disease, the well-known "blue baby" operation, and so on. It is general physiology, however, that is most in his debt. For the notion of the circulating blood is what underlies our present understanding of the self-stabilizing internal environment of the body. In the dynamics of the human system the most important role is played by the fluid whose circulation Harvey discovered by a feat of great insight.

PORTRAIT OF ISAAC NEWTON was painted by Godfrey Kneller in 1689, when Newton was 46. Four years earlier Newton had developed the concept of universal gravitation. Newton's principal work *Philosophiae Naturalis Principia Mathematica* was published in 1687.

Newton's Discovery of Gravity

3

March 1981

*How did he come to develop the concept that marked
the beginning of modern science? In essence he did so
by repititively comparing the real world with a
simplified mathematical representation of it*

The high point of the Scientific Revolution was Isaac Newton's discovery of the law of universal gravitation: All objects attract each other with a force directly proportional to the product of their masses and inversely proportional to the square of their separation. By subsuming under a single mathematical law the chief physical phenomena of the observable universe Newton demonstrated that terrestrial physics and celestial physics are one and the same. In one stroke the concept of universal gravitation revealed the physical significance of Johannes Kepler's three laws of planetary motion, solved the thorny problem of the origin of the tides and accounted for Galileo Galilei's curious and unexplained observation that the descent of a free-falling object is independent of its weight. Newton had achieved Kepler's goal of developing a physics based on causes.

The momentous discovery of universal gravitation, which became the paradigm of successful science, was not the result of an isolated flash of genius; it was the culmination of a series of exercises in problem solving. It was a product not of induction but of logical deductions and transformations of existing ideas. The discovery of universal gravity brings out what I believe is a fundamental characteristic of all great breakthroughs in science from the simplest innovations to the most dramatic revolutions: the creation of something new by the transformation of existing notions.

Newton developed the concept of universal gravity in the first few months of 1685, when he was 42. Physicists have usually made their greatest contributions at a much earlier age, but Newton was still in what he called "the prime years of my life for invention." The documents that have enabled me to date the discovery also make it possible to reconstruct the process that led to it.

A decisive step on the path to universal gravity came in late 1679 and early 1680, when Robert Hooke introduced Newton to a new way of analyzing motion along a curved trajectory. Hooke had cleverly seen that the motion of an orbiting body has two components, an inertial component and a centripetal, or center-seeking, one. The inertial component tends to propel the body in a straight line tangent to the curved path, whereas the centripetal component continuously draws the body away from the inertial straight-line trajectory. In a stable orbit such as that of the moon the two components are matched, so that the moon neither veers away on a tangential path nor spirals toward the earth.

The concept of a centripetal force replaced the older and misleading notion of a centrifugal, or center-fleeing, force. René Descartes and Christiaan Huygens had analyzed curved motion in terms of such a centrifugal force. Descartes, for example, had investigated the movement of a ball on the inner surface of a hollow cylinder and the movement of water in a bucket swung in a circle. The ball and the water seemed to flee the center of the system, and so Descartes attributed their motion to the influence of a centrifugal force. It is now clear there is no such force; a center-fleeing force cannot be traced to the interaction of physical objects. The illusion of a centrifugal force comes about when a moving object is viewed from a rotating frame of reference.

With the change in outlook from centrifugal to centripetal force came an appreciation of the fundamental role of the central body. The centrifugal analysis had focused on the revolving object, whose "endeavor to recede" from the center seems to be independent of the properties of the central body. The concept of a centripetal force, in contrast, depends fundamentally on the central body, toward which the revolving object is impelled or attracted. The interaction of the central, attracting body with the revolving, attracted object can obviously be expected to have a part in any theory of gravitation.

Hooke's analysis of curved motion may seem to be such an obvious and immediate consequence of the Cartesian principle of inertia that Newton would not have needed Hooke to instruct him on the subject as late as 1679. Newton had more or less accepted the inertial principle some 20 years earlier. Nevertheless, Newton, like Descartes and Huygens, was so mired in the concept of centrifugal endeavor that the full implications of inertial physics were far from obvious to him.

On November 24, 1679, Hooke wrote to Newton suggesting that they engage in a private "philosophical" correspondence on scientific topics of mutual interest. Six years earlier they had clashed publicly over Newton's experiments and theories on the prismatic dispersion of light and on the nature of color. Hooke was only one of several investigators who had rejected Newton's optical theories. Newton was so vexed at having to defend his work that he vowed to abandon "philosophy" (physical science) because she was "so litigious a lady" that a man who had anything to do with her would have to spend the rest of his life defending his opinions.

Hooke had since become secretary of the Royal Society of London. In spite of the earlier controversy his letter to Newton was friendly and gracious. The letter invited Newton to comment on any of Hooke's hypotheses or opinions, particularly on the notion of "compounding the celestiall motions of the planetts [out] of a direct motion by the tangent & an attractive motion towards the central body." This sentence was apparently Newton's introduction to the idea of decomposing curved motion into an inertial component and a centripetal one. There is no evidence that he had yet reached Hooke's level of understanding of circular motion. Indeed, Newton still often spoke of orbital motion in terms of centrifugal force.

In his letter Hooke ventured the suggestion that the centripetal force drawing a planet toward the sun varies inversely as the square of the separation. At this point Hooke was stuck. He could not see the dynamical consequences of his own deep insight and therefore could not make the leap from intuitive hunch and guesswork to exact science. He

Mr. WHISTON'S SCHEME of the SOLAR SYSTEM EPITOMIS'D. To w.ᶜʰ is annex'd

A Translation of part of ÿ General *Scholium* at ÿ end of ÿ second Edition of S.ᵗ *Isaac Newton's* Principia. *Concerning God.*

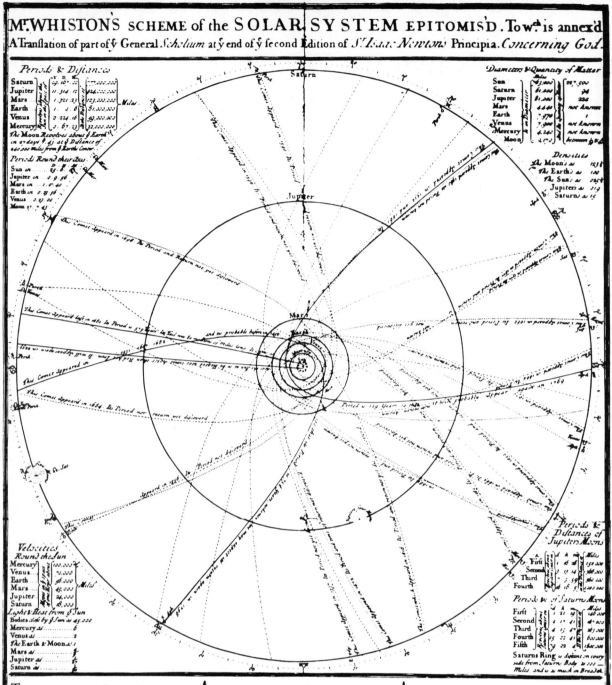

THIS most Elegant System of the Planets and Comets could not be produced but by and under the Contrivance and Dominion of an Intelligent and Powerful Being. And if the Fixed Stars are the Centers of such other Systems, all these being Framed by the like Council will be Subject to the Dominion of *One*, especially seeing the Light of the Fixed Stars is of the same Nature with that of the Sun, and the Light of all these Systems passes mutually from one to another. He governs all things, not as ÿ Soul of the World, but as the Lord of the Universe, and because of his Dominion he is wont to be called Lord God παντοκρατωρ (i.e. Universal Emperor) for God is a Relative word, and has Relation to Servants: And the Deity is the Empire of God, not over his own Body (as is the Opinion of those, who make him the Soul of the World) but over his Servants. The *Supreme God* is a Being Eternal, Infinite, Absolutely Perfect; but a Being however Perfect without Dominion, is not *Lord God*: For we say, *my God, your God, the God of Israel*, but we do not say, *my Eternal, your Eternal, the Eternal of Israel*; we do not say *my Infinite, your Infinite, the Infinite of Israel*; we do not say *my Perfect, your Perfect, the Perfect of Israel*. These Titles have no Relation to Servants. The word God frequently signifies Lord, but every Lord is not God. The Empire of a Spiritual being constitutes God, true

Empire constitutes True God, Supreme the Supreme, Feigned the Feigned. And from his true Empire it follows that the true God is Living, Intelligent & Powerful, from his other Perfections that he is the Supreme or Supremly Perfect. He is *Eternal & Infinite, Omnipotent* and *Omnipresent* that is, he endures from Eternity to Eternity, and he is present from Infinity to Infinity. he Governs all Things, and Knows all Things which are or which can be known. He is not Eternity or Infinity, but he is Eternal and Infinite, he is not Duration or Space, but he Endures and is Present. He endures always and is present everywhere and by existing always and everywhere, he Constitutes Duration and Space, Eternity and Infinity. Whereas every Particle of Space is *always*, and every Individual Moment of Duration is *every where*, certainly the Framer and Lord of the Universe shall not be (*nunquam nusquam*) *never no where*. He is Omnipresent not Virtually only, but also Substantially, for Power without Substance cannot Subsist. In him are contain'd and moved all things (so the Antients thought) but without mutual Passion God suffers nothing from the Motions of Bodies. Nor do they suffer any Resistance from the Omnipresence of God. It is confess'd that the Supreme God exists Necessarily, and by the

same Necessity he is *always* and *every where*. Whence also he is wholly Similar, all Eye, all Ear, all Brain, all Arm, all the Power of Perceiving Understanding and Acting. But after a manner not at all Corporeal, after a manner not like that of Men, after a manner wholly to us unknown. As a Blind Man has no notion of Colours, so neither have we any notion of the manner how the most Wise God perceives and understands all things. He is wholly destitute of all Body and of all Bodily shape, and therefore cannot be seen, heard, nor touched, nor ought to be Worshiped under the Representation of any thing Corporeal. We have Ideas of his Attributes, but we know not at all what is the Substance of any thing whatever. We see only the Figures and Colours of Bodies, we hear only Sounds, we touch only the outward Surfaces, we smell only Odours, and taste Tasts. but we know not by any sense or reflex Act the inward Substances, and much less have we any Notion of the Substance of God. We know him only by his Properties and Attributes and by the most Wise and Excellent Structure of things, and by Final Causes, but we Adore and Worship him on account of his Dominion. For God without Dominion, Providence & Final Causes is nothing else but Fate and Nature.

Engrav'd and Sold by John Senex at the Globe in Salisbury Court near Fleetstreet. Where is to be had D.ʳ Halley's Scheme of the Total Eclipse of the Sun w.ᶜʰ it will be in 1724. Also his Zodiack containing all the Stars in the Way of the Moon and Planets useful in discovering the Longitude at sea

could go no further because he lacked both the mathematical genius of Newton and an appreciation of Kepler's law of areas, which figured prominently in Newton's subsequent approach to celestial dynamics. The law of areas states that the radius vector from the sun to a planet sweeps out equal areas in equal times.

On November 28 Newton wrote to Hooke that before reading Hooke's letter of the 24th he did not "so much as heare (that I remember) of your Hypotheses of compounding the celestial motions of the Planets of a direct motion by the tangent to the curve" and an "attractive" motion toward the sun. Having admitted that Hooke's analysis was new to him, Newton immediately changed the subject to a fancy of his own: the effect of the earth's rotation on a free-falling object. If a dropped object could pass through the rotating earth, what path would the object take? Newton had incorrectly concluded that it would follow a spiral trajectory.

In Hooke's next letter, dated December 9, he caught Newton's error and pointed out that the path "would resemble an Elleipse." Hooke was eager to get Newton going on the problem of planetary motion, and so he suggested that the correct description of an object falling through the earth and his own analysis of planetary motion were both cases of "Circular motions compounded by a Direct motion and an attractive one to a center."

On December 13, 1679, Newton responded guardedly to Hooke's correction but did not comment on his proposed analysis of circular motion. Hooke did not give up. In a letter written on January 6, 1680, he returned to his thesis about curved motion and repeated the quantitative supposition that the centripetal attraction is inversely proportional to the square of the distance. From this supposition Hooke concluded that the velocity of the revolving body is inversely proportional to the distance from the center. He then pointed out that his analysis "doth very Intelligibly and truly make out all the Appearances of the Heavens." Newton did not reply.

On January 17 Hooke sent a short supplementary letter in which he wrote: "It now remaines to know the proprietys of a curve Line (not circular nor concentricall) made by a centrall attractive

power which makes the velocitys of Descent from the tangent Line or equall straight motion at all Distances in a Duplicate proportion reciprocally taken." In modern terminology Hooke's problem can be paraphrased as follows: If a central attractive force causes an object to fall away from its inertial path and move in a curve, what kind of curve results if the attractive force varies inversely as the square of the distance? He concluded: "I doubt not but that by your excellent method you will easily find out what that Curve must be, and its proprietys, and suggest a physicall Reason of this proportion."

Newton evidently did do almost that. He proved that an ellipse would satisfy the conditions outlined by Hooke. Nevertheless, he did not communicate the result of this proof to Hooke or to anyone else until August, 1684, when he was visited by Edmund Halley, the astronomer and mathematician. Halley came to see Newton in order to ask "what he thought the Curve would be that would be described by the Planets, supposing the force of attraction towards the Sun to be reciprocal to the square of their distance from it." The problem had been much discussed by the Royal Society. Halley and Christopher Wren were unable to solve it, and Hooke never produced a solution, although he maintained he had found one.

When Newton heard the question, he responded immediately: an ellipse. Halley asked him how he knew and Newton replied: "I have calculated it." Newton apparently could not find the calculations, but at Halley's urging he wrote them up for the Royal Society in the small tract De Motu (Concerning Motion). In De Motu Newton described his work on terrestrial and celestial dynamics, including his ideas on motion in free space and in a resistive medium. Newton must have finished De Motu by December 10, 1684, because Halley told the Royal Society then that Newton had recently shown him the curious treatise.

The exact progression of Newton's ideas in the time between his correspondence with Hooke and his completion of the first draft of De Motu is not documented. Nevertheless, I am certain it was Hooke's method of analyzing curved motion that set Newton on the right track. Although not all historians would agree with me, I believe the approach Newton takes to terrestrial and

celestial dynamics in De Motu, which he further developed the following spring in the first book of the Philosophiae Naturalis Principia Mathematica, represents his thinking on planetary dynamics inspired by his correspondence with Hooke. In a few autobiographical manuscripts Newton said the correspondence either preceded or coincided with his demonstration published first in De Motu and then in the Principia that an object that has an inertial motion and is subject to an inverse-square centripetal force moves in an elliptical orbit.

It was this demonstration that brought out the physical significance of Kepler's law of elliptical orbits (the law stating that each planet moves in an elliptical path with the sun at one focus of the ellipse). The modern reader may be surprised that it was not Kepler but Newton who revealed the fundamental nature of Kepler's laws of planetary motion. Before the publication of the Principia, however, these laws (which were even called hypotheses) were not as highly respected as they came to be afterward.

Kepler's law of areas in particular had a diminished status in the 17th century. Most astronomical works did not even mention it. For example, Thomas Streete's Astronomia Carolina, from which Newton copied Kepler's third law (The cube of the average distance of a planet from the sun is proportional to the square of the orbital period), never discusses the law of areas or hints at its existence. Most 17th-century astronomers calculated planetary positions not by the law of areas but by a construction based on a uniformly rotating vector emanating from the empty focus of the planet's elliptical orbit [see top illustration on page 22]. Since astronomers rarely employed the law of areas, it required extraordinary perception for Newton to see its significance. Newton was the one who elevated Kepler's law of areas to the status it enjoys today.

The very first proposition of the Principia (and the discussion at the beginning of De Motu) develops the dynamical significance of the law of areas by proving that the curved motion described by the law is a consequence of centripetal force. The proof, which has three parts, shows how well Newton had learned Hooke's technique of decomposing curved motion into an inertial component and a centripetal one.

In the first part of the proof Newton considers a body moving along a straight line with a constant velocity. The line is divided into equal intervals to indicate that the body moves equal distances in equal times. A point P is chosen at a distance h above the line of motion. The triangles formed by connecting P to any of the equal intervals all have the same area because they have equal bases and the same altitude h. By this simple analysis Newton revealed

NEWTONIAN SYSTEM OF THE WORLD was diagrammed by William Whiston, who succeeded Newton as Lucasian Professor at the University of Cambridge. The diagram is from Whiston's broadside "Scheme of the Solar System Epitomis'd," published in 1724. The planets and the satellites of Jupiter and Saturn are shown orbiting the sun under the action of universal gravity. Remarkably, Whiston also included the orbits of comets. Newton had shown that orbits of comets are ellipses or parabolas in which a vector from the sun to the comet sweeps out equal areas in equal times. Below the diagram is Whiston's translation of part of the final General Scholium of the Principia (which is from the second edition, published in 1713). There Newton wrote that "This most Elegant System of the Planets and Comets could not be produced but by and under the Contrivance and Dominion of an Intelligent and Powerful Being."

an unexpected relation between inertial motion and the law of areas.

In the second part of the proof the body moves as before initially, but at the end of the second interval it receives an impulsive force—a blow—toward P. Therefore in the third interval the body no longer moves along the original straight line but rather along another straight line closer to P. Newton again showed by geometry that the triangle formed by connecting P to the ends of the trajectory traced in the second interval has the same area as the triangle formed by connecting P to the ends of the trajectory traced in the third interval.

In the third part the body is given a blow toward P at the end of each interval. As a result the body moves in a polygonal path around P. Again the area relation holds. In the limiting case where the interval between blows approaches zero the body is subject to a continuous force directed toward P and the polygonal path becomes a smooth curve or orbit. In this way Newton proved that a centripetal force generates a curve according to the law of areas.

The second proposition of the *Principia* proves the converse: Motion in a curve described by the law of areas implies a centripetal force. With these two propositions Newton demonstrated that the law of areas is a necessary and sufficient condition for inertial motion in a central-force field.

The two propositions are part of a sequence of demonstrations that begins with the law of areas and ends with a proof that an elliptical orbit requires an inverse-square centripetal force. This sequence of demonstrations, presented both in the *Principia* and in *De Motu*, marks a profound discontinuity in the history of the exact sciences. The demonstrations introduced a radically new celestial dynamics based on new concepts of force, momentum, mass and inertia and a wholly novel quantitative measure of dynamical force. The subtitle of Kepler's *Astronomia Nova* set the goal of creating a "celestial physics based on causes." Newton achieved this goal, of which Kepler had had only a visionary glimpse. Neither Galileo nor Descartes had conceived of such a celestial dynamics. And the Newtonian formulation left even the great physicist Huygens far behind.

From the early draft of *De Motu* that Newton probably wrote in November, 1684, it is clear he had not yet developed the concept of universal gravitation. The draft discusses centripetal force directed toward the focus of an ellipse and concludes with the scholium, "Therefore the major planets revolve in ellipses having a focus in the center of the sun, and radii drawn [from the planets] to the sun describe areas proportional to the times, entirely as Kepler supposed...."

Newton neither proved this scholium nor continued to believe it for long, and strictly speaking it is false. As he soon realized, the planets do not move according to the law of areas in simple Keplerian elliptical orbits with the sun at a focus. Instead the focus lies in the common center of mass because not only does the sun attract each planet but also each planet attracts the sun (and the planets attract one another). If Newton had already formulated his principle of universal gravitation, he would not have proposed the erroneous scholium.

Newton quickly realized he had not proved that the planets move precisely according to the law of elliptical orbits and the law of areas. He had only found that the laws hold for a one-body system: a single point mass moving with an initial component of inertial motion in a central-force field. He recognized that the one-body system corresponds not to the real world but to an artificial situation that is easier to investigate mathematically. The one-body system reduces the earth to a point mass and the sun to an immobile center of force.

LETTER TO NEWTON from Robert Hooke includes Hooke's views on the analysis of motion along a curved trajectory. (The letter is dated January 6, 1679, according to a version of the Julian calendrical system in which the year started in March; the modern calendrical system puts the date at January 6, 1680.) In the second sentence Hooke proposes that "the Attraction is always in a duplicate proportion to the Distance from the Center Reciprocall" (that is, the attraction is inversely proportional to the distance squared). As a result "the Velocity will be in a subduplicate proportion to the Attraction and consequently as Kepler supposes Reciprocall to the Distance." Hooke states that this analysis explains "all the Appearances of the Heavens." He stresses the importance of "finding out the proprietys" of curves because longitudes, which are "of great Concerne to Mankind," can be derived from the moon's curved motion.

What enabled Newton to transcend the one-body system was his appreciation of the consequences of his third law of motion: the law of action and reaction. This law is perhaps the most original of his three laws of motion (the other two are the law of inertia and the force law). One testimonial to its novelty is that even today it is often employed incorrectly by those who relate it not to an impact situation or to the interaction of bodies but to a supposed condition of equilibrium.

The development of Newton's thinking on action and reaction after he completed the first draft of *De Motu* is set out in the opening sections of the first book of the *Principia*. In the introduction to the 11th section Newton explains that he has confined himself so far to a situation that "hardly exists in the real world," namely the "motions of bodies attracted toward an unmoving center." The situation is artificial because "attractions customarily are directed toward bodies and—by the third law of motion—the actions of attracting and attracted bodies are always mutual and equal." As a result, "if there are two bodies, neither the attracting nor the attracted body can be at rest." Rather, "both bodies (by the fourth corollary of the laws) revolve about a common center, as if by a mutual attraction."

Newton had seen that if the sun pulls on the earth, the earth must also pull on the sun with a force of equal magnitude. In this two-body system the earth does not move in a simple orbit around the sun. Instead the sun and the earth each move about their mutual center of gravity. A further consequence of the third law of motion is that each planet is a center of attractive force as well as an attracted body; it follows that a planet not only attracts and is attracted by the sun but also attracts and is attracted by each of the other planets. Here Newton has taken the momentous step from an interactive two-body system to an interactive many-body system.

In December, 1684, Newton completed a revised draft of *De Motu* that describes planetary motion in the context of an interactive many-body system. Unlike the earlier draft the revised one concludes that "the planets neither move exactly in ellipses nor revolve twice in the same orbit." This conclusion led Newton to the following result: "There are as many orbits to a planet as it has revolutions, as in the motion of the Moon, and the orbit of any one planet depends on the combined motion of all the planets, not to mention the actions of all these on each other." He then wrote: "To consider simultaneously all these causes of motion and to define these motions by exact laws allowing of convenient calculation exceeds, unless I am mistaken, the force of the entire human intellect."

There are no documents that indicate

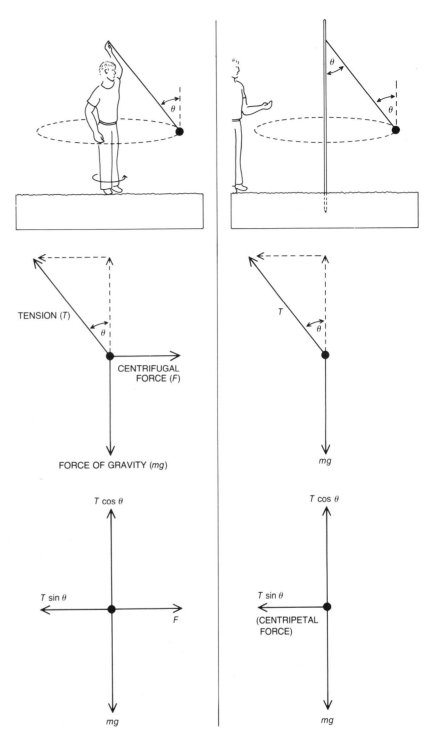

CENTRIFUGAL FORCE is a fictitious force. The illusion of a centrifugal, or center-fleeing, force can arise when a moving object is viewed from a rotating frame of reference (*left*), as when a ball is swung at the end of a string by an observer who rotates with the same angular speed as the ball. Two known forces act on the ball: the tension of the string and the force of gravity. The ball is not accelerating in the vertical direction, and so all vertical forces acting on it must be in balance; in particular the vertical component of the tension cancels the force of gravity. Since the observer and the ball are rotating together, the ball appears to be at rest and it seems that the horizontal forces should also be in balance. As a result the observer postulates a centrifugal force that cancels the horizontal component of the tension. No such force, however, can be traced to the interaction of physical objects. A different analysis of forces results (*right*) when the ball is rotating in the same way but the observer is at rest. In this stationary frame of reference the observer sees the same vertical forces on the ball as he saw in the rotating frame. In the horizontal direction, however, the ball is not at rest with respect to the observer but is moving in a circle. In other words, the ball accelerates continuously toward the center, so that the horizontal forces should not be expected to balance. The ball is subject to a centripetal, or center-seeking, force which is the horizontal component of the tension of the string. The centripetal force can be traced to the interaction of two physical objects: the string and the ball.

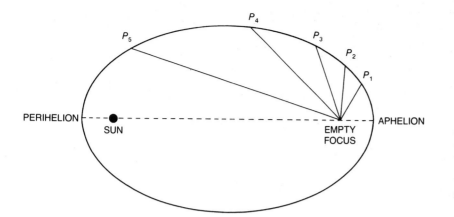

PERIHELION

SUN

EMPTY FOCUS

APHELION

PLANETARY POSITIONS were often found in the 17th century not by Kepler's law of areas but by a construction based on a uniformly rotating radius vector that emanates from the empty focus of a planet's elliptical orbit. The position of a planet (P_1, P_2, P_3, P_4, P_5) at successive moments is the intersection of the ellipse and the vector. Kepler's law of areas states that the radius vector from the sun to a planet sweeps out equal areas in equal times. As a result the planet moves slower at aphelion than at perihelion. The diagrammed construction gives the same result. Correction factors were added to make the construction fit the data more accurately.

how, in the month or so between writing the first draft of *De Motu* and revising it, Newton came to perceive that the planets act gravitationally on one another. Nevertheless, the passage cited above expresses this perception in unambiguous language: "*eorum omnium actiones in se invicem*" ("the actions of all these on each other"). A consequence of this mutual gravitational attraction is that all three of Kepler's laws are not strictly true in the world of physics but are true only for a mathematical construct in which point masses that do not interact with one another orbit either a mathematical center of force or a stationary

attracting body. The distinction Newton draws between the realm of mathematics, in which Kepler's laws are truly laws, and the realm of physics, in which they are only "hypotheses," or approximations, is one of the revolutionary features of Newtonian celestial dynamics.

I have assumed that the third law of motion was the key factor in the reasoning that led Newton to suggest mutual gravitational perturbations of planetary orbits. There is no direct evidence for my assumption because no documents exist in which there is an antecedent version of his statement "the actions

of all these on each other." Nevertheless, there is strong indirect evidence. In the spring of 1685, a few months after revising *De Motu*, Newton was well on his way to finishing the first draft of the *Principia*. In the initial version of what was to become a second book, "The System of the World," he spelled out the steps that led him to the concept of planetary gravitational interactions. In these steps the third law of motion has the chief role, and I see no reason to believe they are not the same steps that led him to the same concept a few months earlier when he revised *De Motu*.

Here are two passages from the first draft of "The System of the World" (translated from the Latin by Anne Whitman and me) that bring out the crucial role of the third law of motion:

"20. *The agreement between the analogies.*

"And since the action of centripetal force upon the attracted body, at equal distances, is proportional to the matter in this body, it is reasonable, too, that it is also proportional to the matter in the attracting body. For the action is mutual, and causes the bodies by a mutual endeavor (by law 3) to approach each other, and accordingly it ought to be similar to itself in both bodies. One body can be considered as attracting and the other as attracted, but this distinction is more mathematical than natural. The attraction is really that of either of the two bodies toward the other, and thus is of the same kind in each of the bodies.

"21. *And their coincidence.*

"And hence it is that the attractive force is found in both bodies. The sun attracts Jupiter and the other planets,

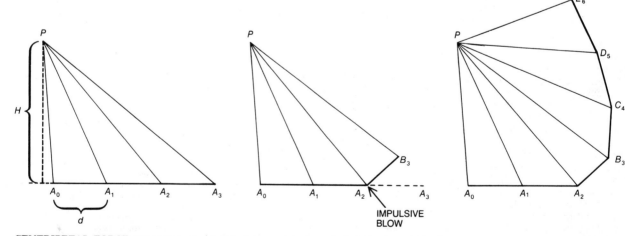

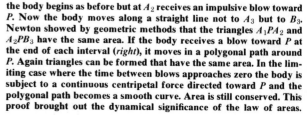

CENTRIPETAL FORCE generates a curved trajectory consistent with the law of areas. This property of a centripetal force was demonstrated by Newton in the first proposition of the *Principia* and in the discussion at the beginning of short tract *De Motu* (*Concerning Motion*). Newton began (*left*) by considering a body moving in a straight line at a constant speed. The body starts at A_0 and after successive equal intervals reaches first A_1, then A_2 and so on. A point P is chosen above the line of motion. The triangles A_0PA_1, A_1PA_2, A_2PA_3 and so forth all have the same area because they have equal bases and the same altitude. In a second stage of the analysis (*middle*)

the body begins as before but at A_2 receives an impulsive blow toward P. Now the body moves along a straight line not to A_3 but to B_3. Newton showed by geometric methods that the triangles A_1PA_2 and A_2PB_3 have the same area. If the body receives a blow toward P at the end of each interval (*right*), it moves in a polygonal path around P. Again triangles can be formed that have the same area. In the limiting case where the time between blows approaches zero the body is subject to a continuous centripetal force directed toward P and the polygonal path becomes a smooth curve. Area is still conserved. This proof brought out the dynamical significance of the law of areas.

Jupiter attracts its satellites and similarly the satellites act on one another and on Jupiter, and all the planets on one another. And although the actions of each of a pair of planets on the other can be distinguished from each other and can be considered as two actions by which each attracts the other, yet inasmuch as they are between the same two bodies they are not two but a simple operation between two termini. Two bodies can be drawn to each other by the contraction of one rope between them. The cause of the action is twofold, namely the disposition of each of the two bodies; the action is likewise twofold, insofar as it is upon two bodies; but insofar as it is between two bodies it is single and one. There is not, for example, one operation by which the sun attracts Jupiter and another operation by which Jupiter attracts the sun, but one operation by which the sun and Jupiter endeavor to approach each other. By the action by which the sun attracts Jupiter, Jupiter and the sun endeavor to approach each other (by law 3), and by the action by which Jupiter attracts the sun, Jupiter and the sun also endeavor to approach each other. Moreover, the sun is not attracted by a twofold action toward Jupiter, nor is Jupiter attracted by a twofold action toward the sun, but there is one action between them by which both approach each other."

Next Newton concluded that "according to this law all bodies must attract each other." He proudly presented the conclusion and explained why the magnitude of the attractive force is so small that it is unobservable. "It is possible," he wrote, "to observe these forces only in the huge bodies of the planets."

In book three of the *Principia,* which is also concerned with the system of the world but is somewhat more mathematical, Newton treats the topic of gravitation in essentially the same way. First, in what is called the moon test, he extends the weight force, or terrestrial gravity, to the moon and demonstrates that the force varies inversely with the square of the distance. Then he identifies the same terrestrial force with the force of the sun on the planets and the force of a planet on its satellites. All these forces he now calls gravity. With the aid of the third law of motion he transforms the concept of a solar force on the planets into the concept of a mutual force between the sun and the planets. Similarly, he transforms the concept of a planetary force on the satellites into the concept of a mutual force between planets and their satellites and between satellites. The final transformation is the notion that all bodies interact gravitationally.

M y analysis of the stages of Newton's thinking should not be taken as diminishing the extraordinary force of his creative genius; rather, it should make that genius plausible. The analy-

PAGE FROM A DRAFT OF "DE MOTU" that Newton probably wrote in November, 1684, is in his handwriting. In *De Motu* Newton discussed terrestrial and celestial dynamics, including the idea of centripetal force directed toward the focus of an ellipse. The page ends with the scholium, "Therefore the major planets revolve in ellipses having a focus in the center of the sun, and radii drawn [from the planets] to the sun describe areas proportional to the times, entirely as Kepler supposed...." The scholium is false, and the nature of the error indicates that Newton had not yet developed the concept of universal gravitation. As Newton soon realized, the focus of the orbits of the planets is not the sun but the center of mass common to the planets and the sun. Not only does the sun attract each planet but also each planet attracts the sun.

sis shows Newton's fecund way of thinking about physics, in which mathematics is applied to the external world as it is revealed by experiment and critical observation. This way of thinking, which I call the Newtonian style, is captured by the English title of Newton's great work: *Mathematical Principles of Natural Philosophy.*

The Newtonian style consists in a repeated give-and-take between a mathematical construct and physical reality. In the development of Newton's ideas on gravity and in his presentation of those ideas in the *Principia,* he started with a mathematical construct that represents nature simplified: a point mass moving around a center of force. Because he did not assume that the construct was an exact representation of the physical universe he was free to explore the properties and effects of a mathematical attractive force even though he found the concept of a grasping force "acting at a distance" to be abhorrent and not admissible in the realm of good physics. Next he compared the consequences of his mathematical construct with the observed principles and laws of the external world such as Kepler's law of areas and law of elliptical orbits. Where the mathematical construct fell short Newton modified it. He made the center of force not a mathematical entity but a point mass. I say a point mass rather than a physical body because he had not yet considered physical properties such as size, shape and mass.

From the modified mathematical construct Newton concluded that a set of point masses circling the central point mass attract one another and perturb one another's orbits. Again he compared the construct with the physical world. Of all the planets, Jupiter and Saturn are the most massive, and so he sought orbital perturbations in their motions. With the help of John Flamsteed, Newton found that the orbital motion of Saturn is perturbed when the two planets are closest together. The process of repeatedly comparing the mathematical construct with reality and then suitably modifying it led eventually to the treatment of the planets as physical bodies with definite shapes and sizes.

After Newton had modified the construct many times he applied it to the system of the world. He asserted that the force of attraction, which he had derived mathematically, is universal gravity. He found that the moon moves as if it were attracted to the earth with a force that is 1/3,600th of the strength of the gravitational force with which the earth pulls on objects at its surface. Since the moon is 60 times farther from the center

of the earth than objects on the earth's surface are, the factor of 1/3,600 is consistent with the deduction that the earth's gravity extends to the moon and diminishes with the square of the distance.

The law of universal gravitation explains why the planets follow Kepler's laws approximately and why they depart from the laws in the way they do. It demonstrates why (in the absence of friction) all bodies fall at the same rate at any given place on the earth and why the rate varies with elevation and latitude. The law of gravitation also explains the regular and irregular motions of the moon, provides a physical basis for understanding and predicting tidal phenomena and shows how the earth's rate of precession, which had long been observed but not explained, is the effect of the moon's pulling on the earth's equatorial bulge. Since the mathematical force of attraction works well in explaining and predicting the observed phenomena of the world, Newton decided that the force must "truly exist" even though the received philosophy to which he adhered did not and could not allow such a force to be part of a system of nature. And so he called for an inquiry into how the effects of universal gravity might arise.

Although Newton at times thought universal gravity might be caused by the impulses of a stream of ether particles bombarding an object or by variations in an all-pervading ether, he did not advance either of these notions in the *Principia* because, as he said, he would "not feign hypotheses" as physical explanations. The Newtonian style had led him to a mathematical concept of universal force, and that style led him to apply his mathematical result to the physical world even though it was not the kind of force in which he could believe.

Some of Newton's contemporaries were so troubled by the idea of an at-

tractive force acting at a distance that they could not begin to explore its properties, and they found it difficult to accept the Newtonian physics. They could not go along with Newton when he said he had not been able to explain how gravity works but that "it is enough that gravity really exists and suffices to explain the phenomena of the heavens and the tides." Those who accepted the Newtonian style fleshed out the law of universal gravity, showed how it explained many other physical phenomena and demanded that an explanation be sought of how such a force could be transmitted over vast distances through apparently empty space. The Newtonian style enabled Newton to study universal gravity without premature inhibitions that would have blocked his great discovery. The 18th-century biologist Georges Louis Leclerc de Buffon once wrote that a man's style cannot be distinguished from the man himself. In the case of Newton his greatest discovery cannot be separated from his style.

The correspondence between Hooke and Newton clearly shows that Hooke taught Newton how to analyze curved motion. Hooke subsequently made the much stronger claim that he deserved credit for suggesting to Newton the law of universal gravity varying inversely with the square of the distance. Many historians have echoed Hooke's view.

The claim, however, does not hold up. Hooke had merely suggested that the planets are subject to an inverse-square force directed toward the sun. Universal gravitation is much more than a solar-directed force. It also implies an effect of the planets on the sun. What is more, it applies to all objects in the universe. The law of universal gravitation is not merely an inverse-square relation; it is also a mathematical relation between the masses of the attracting bodies. It took tremendous insight to leap from an inverse-square solar-directed force to universal gravitation. And it took the genius of Newton to invent the modern concept of mass.

Newton did not feel he owed Hooke credit even for suggesting that the centripetal force is inversely proportional to the square of the distance. In 1673 Huygens had published a supplement to a book on the pendulum clock in which he states that for circular motion the centrifugal force is measured by v^2/r, where v is the velocity of the orbiting body and r is the radius of rotation. Newton had independently discovered the same relation in the 1660's. Since the mathematical difference between a centrifugal force and a centripetal force is only a matter of direction, the v^2/r relation also holds for a centripetal force. From this relation and Kepler's third law it follows by simple algebra that the centripetal force varies inversely with

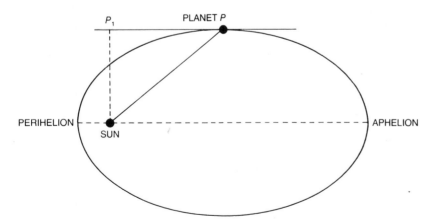

ORBITAL SPEED OF A PLANET is inversely proportional not to the direct distance between the sun and the planet but to the perpendicular distance (the distance represented by the broken line between the sun and the tangent to the orbit *PP'*). Only at two points in the orbit (perihelion and aphelion) are the direct distance and the perpendicular distance the same.

the square of the distance. After Huygens' book was published anyone with a rudimentary knowledge of algebra could have found an inverse-square centripetal force for a circular orbit. Accordingly Newton saw no need to acknowledge Hooke's statement of an inverse-square law.

Both Hooke and Newton were aware that finding an inverse-square law for circular orbits was not the same thing as showing that the law holds for elliptical orbits in which the motion follows Kepler's law of areas. The task, which Newton carried out, was to demonstrate that an inverse-square law of centripetal force corresponds to orbital motion according to Kepler's law of elliptical orbits and his law of areas. In discussing this point in the letter dated January 6, 1680, Hooke made a fundamental error that must have convinced Newton that Hooke did not entirely understand what

he was talking about. Hooke said that if the attraction varies inversely as the square of the distance, the orbital speed of a planet will be "as Kepler supposes Reciprocall to the Distance." Yet under the conditions Hooke assumed the orbital speed is not inversely proportional to the direct distance from the sun except at the extreme points of the orbit: perihelion and aphelion. In view of Hooke's error Newton was not about to give him credit for having suggested the inverse-square nature of the centripetal force.

In 1717 Newton wanted to ensure his own priority in discovering the inverse-square law of gravitation, and so he invented a scenario in which he made the famous moon test not while writing the *Principia* but two decades earlier in the 1660's. The documents of the 1660's, however, indicate that he was not then comparing the falling of the moon in its

orbit with the falling of objects on the earth but was comparing the "centrifugal endeavor" of the orbiting moon with the "centrifugal endeavor" of a body on the earth's surface rotating along with the daily motion of the earth. He did calculate that for circular planetary orbits the "centrifugal endeavor" would be inversely proportional to the planet's distance from the sun, but he drew no physical conclusions from the calculation.

Newton never published his invented scenario of the early moon test. He included it in the manuscript draft of a letter to the French writer Pierre Des Maizeaux but then crossed it out. Newton also circulated the familiar story that a falling apple set him on a chain of reflections that led to the discovery of universal gravitation. Presumably this invention was also part of his campaign to push back the discovery of gravity, or at least the roots of the discovery, to a time 20 years before the *Principia*.

The real roots of the discovery cannot be put any earlier than December, 1684, when Newton first recognized that if the sun attracts the earth, the earth must attract the sun with a force of equal magnitude. In 1685 he overcame his usual reluctance to write up his discoveries and started to draft the *Principia* for publication by the Royal Society. Perhaps his willingness to present his work for public inspection (and thereby risk possible disapprobation) was motivated first by his momentous discovery of interplanetary perturbations followed by his bold conception of universal gravity. He had within his grasp the foundation of a new system of natural philosophy that could be expounded on mathematical principles. In short, once Newton had something of real consequence to say about celestial dynamics he was willing and even eager to present it to the world.

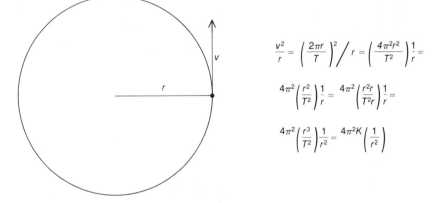

$$\frac{v^2}{r} = \left(\frac{2\pi r}{T}\right)^2 \bigg/ r = \left(\frac{4\pi^2 r^2}{T^2}\right)\frac{1}{r} =$$

$$4\pi^2\left(\frac{r^2}{T^2}\right)\frac{1}{r} = 4\pi^2\left(\frac{r^2 r}{T^2 r}\right)\frac{1}{r} =$$

$$4\pi^2\left(\frac{r^3}{T^2}\right)\frac{1}{r^2} = 4\pi^2 K\left(\frac{1}{r^2}\right)$$

INVERSE-SQUARE NATURE OF CENTRIPETAL FORCE for circular orbits can be deduced from Kepler's third law of planetary motion and from the law of centripetal force. According to Kepler's third law, r^3/T^2 is a constant K, where r is the radius of the planet's orbit and T is the period of the orbit. The law of centripetal force states that for a circular orbit the centripetal force is v^2/r, where v is the planet's velocity. In time T the planet makes a complete orbit, moving a distance $2\pi r$ (the circumference of a circle), and so the velocity is $2\pi r/T$.

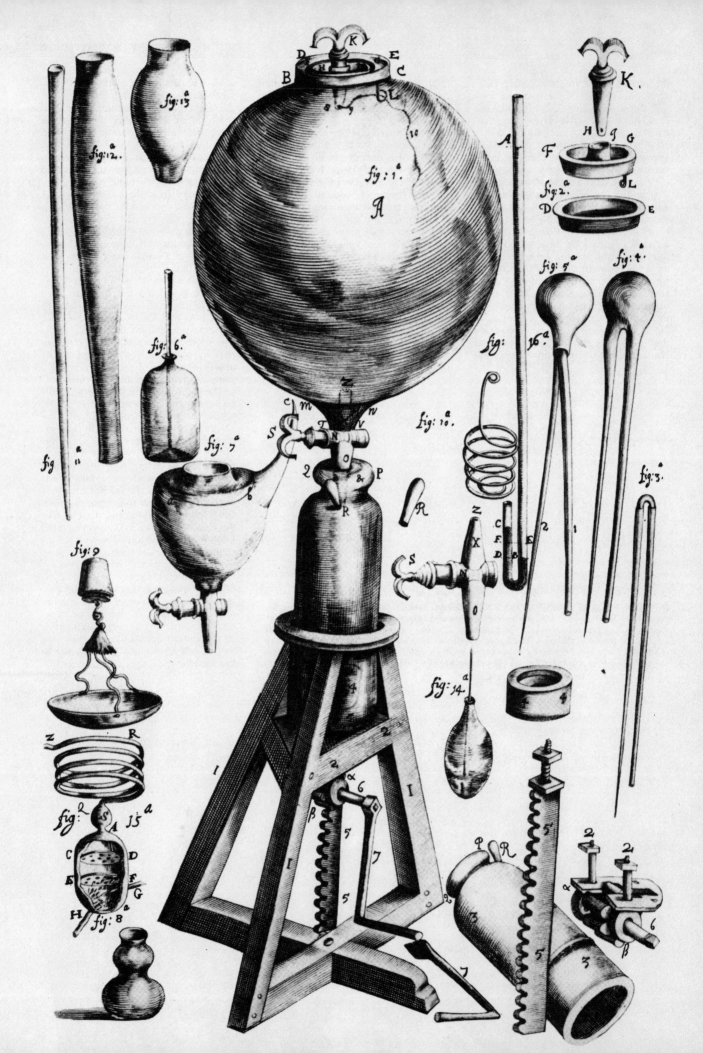

Robert Boyle **4**

by Marie Boas Hall
August 1967

He stands out among 17th-century natural philosophers as a careful observer and investigator. Particularly in pneumatics and chemistry, his experiments laid a firm groundwork for much that was to follow

When Robert Boyle died in 1691, Christian Huygens and Gottfried Wilhelm von Leibniz commiserated that he had wasted his talents trying to prove by experiments what they knew to be true in the light of reason—that he was more interested in observation than in reasoning and had left no unified body of thought. These great contemporaries of Boyle's misunderstood his purpose and aim. With his discoveries in pneumatics, his investigations in optics and physiology and particularly his shaping of the emerging science of chemistry, he had established himself as the leading proponent in England of the new "experimental philosophy." The Cartesian philosophers on the Continent could not perceive that his reputation was the more secure because his discoveries and his theoretical concepts, being grounded in experiment, would provide a firm basis for further work. Unlike the purely rational systems built by so many of his predecessors and contemporaries, his contribution was not to be torn down and superseded.

Robert Boyle was born into an outwardly secure world in 1627, the son of an Elizabethan adventurer who had carved for himself a large property in

BOYLE'S AIR PUMP consisted of a 30-quart glass "receiver" (*A*) connected by a stopcock (*N*) to a 14-inch-long brass pumping cylinder through which a padded piston (*4*) was drawn by a toothed shaft (*5*). First the stopcock was closed and the piston was cranked down. Then, with the stopcock opened, Boyle wrote, "part of the Air formerly contain'd in the Receiver, will nimbly descend into the Cylinder." Then the stopcock was closed, the brass plug (*R*) removed and the piston raised, expelling air from the cylinder. As the procedure was repeated, the air in the receiver was attenuated. Other objects shown are accessories for experiments.

Ireland. As a boy he was sent to be educated on the Continent, where he traveled for several years learning languages, a European outlook and a little about the new trends and problems of science. When he returned to England in 1644 at the age of 17 his father was dead and his older brothers were fighting rebellion in Ireland. During the English Civil War his family was divided in allegiance, some members remaining loyal to the king and some supporting the Cromwellians—in both cases temperately, so that none of them lost his life or his wealth.

Boyle's own preference was to avoid public affairs. Guided by his brother Roger and his elder sister Lady Ranelagh, who was a friend of John Milton's and other Puritan intellectuals, he tried his hand at literature and then turned to agriculture and medicine. Medicine led him to take up chemistry; he found it an enthralling pastime and soon became proficient in it. He was not, however, to become the conventional 17th-century chemist. Although he read widely in chemical and medical works, he was saved from the mysticism of contemporary chemists by his earlier acquaintance with contemporary physics. He had read Francis Bacon and Galileo as a boy. Descartes's *Principles of Philosophy*, published in 1644, influenced his later thought far more deeply than the cloudy views of alchemists did. The fascination of chemical experiment did not keep him from studying the novel ideas in physics and astronomy that were then characterized as the "new learning"; he thought as much about the physical nature of air and Evangelista Torricelli's recent experiments with a column of mercury as he did about the best ways of compounding drugs, and he was as familiar with Copernican astronomy as he was with conflicting interpretations of alchemical doctrine.

By 1653 Boyle was beginning to know some of the most original and active English scientists and to find his aims were theirs. Since 1645 there had been weekly meetings in London, often at Gresham College, of physicians, natural philosophers and mathematicians to discuss the "new learning" they attributed to Bacon and Galileo. The subject matter was broad: the circulation of the blood and the Copernican hypothesis; comets, Jupiter's satellites and the valves in the veins; the nature of air, the appearance of the moon, the improvement of telescopes and the possibility of creating a vacuum; sunspots and novas and the fall of heavy bodies.

By 1650 several members of this group had moved to Oxford, where they continued to meet and where Boyle was invited to join them. He did so in 1654, settling in Oxford for 14 years. The London meetings continued, and after the Restoration they were attended by Royalists who returned to the capital and by many of the Oxford group. It was after a Gresham College lecture on November 28, 1660, that 12 men, including Boyle, decided to establish a "college" to promote "Physico-Mathematicall, Experimentall Learning." That was the beginning of the Royal Society, chartered in 1662 as the first scientific society in Britain and one of the first in the world. Boyle was to become one of its most active members and proudest ornaments.

The scientific circle he joined at Oxford in 1654 was a brilliant one from which he had much to learn. It included England's leading mathematician, John Wallis; Cromwell's brother-in-law John Wilkins, an early adherent of Galileo and a great exponent of Copernicanism; Seth Ward, an early follower of Kepler who had his own original theories in astronomy; distinguished physicians, such as Jonathan Goddard, who were in-

PORTRAIT OF BOYLE was engraved from a drawing made by William Faithorne about 1664. Boyle, then living in Oxford, was already noted for his air pump (*right background*).

miliarity because the writings of Epicurus were well known (Boyle, for instance, had read them as a boy).

By the middle 1650's Boyle had worked out his own version of the mechanical philosophy—the "corpuscular philosophy," he called it—in which he drew on both the Cartesian and the atomic views but wholly accepted neither. He believed "those two grand and most Catholic principles, matter and motion," sufficed to explain all the properties of matter as we experience it. He rejected the Epicurean idea that the specific shapes of atoms determined the nature of substances and also Descartes's concept of an undetectable "aether"; he thought Descartes was wrong in rejecting the possibility of the vacuum and proved the Epicureans were wrong in insisting that there were "frigorific atoms," special corpuscles that produced freezing. Above all, he believed it should be possible to demonstrate the truth of the mechanical philosophy by experiment— particularly by chemical experiments, in which, he held, the scientist was manifestly dealing with small, discrete particles of matter that could survive chemical transmutations and so must have a real and definite existence.

The notion that one could *prove* a scientific theory by experiment was itself novel, and it was by no means universally acceptable to 17th-century thinkers. Boyle (and Isaac Newton after him) saw clearly just how a theory might stand or fall in the light of experimental evidence. Many others—Huygens and Hooke among them—followed Descartes in the belief that the ultimate test of a theory was the appeal to reason. In 1662 and 1663 a notable exchange of views between Boyle and the leading contemporary rationalist philosopher, Benedict Spinoza, was conducted in a correspondence between Spinoza and Henry Oldenburg, secretary of the Royal Society and Boyle's literary assistant. Spinoza could not understand how Boyle could subordinate reason to experiment, or indeed how a chemical combination of particles could behave differently from a physical mixture. The correspondence provides a fascinating example of how one of the most powerful intellects of his day totally failed to grasp concepts that now appear to be obvious. It is also evidence for the originality of Boyle's approach to scientific proof—and to chemistry.

Boyle's major contributions in chemistry were not to be published for some time because in 1657 his attention was turned from chemical to physical

terested in the new chemical remedies; bright young undergraduates and beginners such as Christopher Wren and Robert Hooke, the latter to become Boyle's assistant and later a distinguished natural philosopher in his own right.

Boyle was accepted as being fully competent to share the group's interests, whether mathematical, physical or biological. Curiously he suffered some ridicule for his devotion to chemistry. Whereas the study of chemistry seemed reasonable enough in relation to medicine, it was less understandable in someone, as Boyle himself put it, "who professed to cure no disease but that of ignorance."

Even in the 1650's Boyle was deeply interested in chemistry as the key to the nature of matter. He had begun to compile material for a dozen or more books on the "mechanical philosophy," which was coming to be the most commonly accepted theory of matter. It rejected the long-established Aristotelian

theory of matter in favor of one based on new principles of mechanics established by Galileo and Descartes. Descartes in particular had shown in his *Principles of Philosophy* how, by defining matter and establishing the natural laws of motion, one could construct an entire system of the world without recourse to the vague world of Aristotelian "forms" and "qualities." Heat, for example, instead of being a "form" inherent in substance (and so not further capable of definition) could be regarded as arising from the size, shape and motion of the particles of matter.

The English scientists were much influenced by Descartes's careful formulation of his mechanical philosophy, toward which they were further predisposed by their adherence to similar ideas of Bacon's. Some preferred Pierre Gassendi's modernized and Christianized version of Greek atomic thought as it had been developed by Epicurus; Gassendi's work had appeared several years after Descartes's but it had the appeal of fa-

experiment by news of an exciting new instrument: the air pump invented by Otto von Guericke, who had utilized an ordinary suction pump to empty first a wine barrel filled with water and then a metal receptacle containing only air. Boyle first learned of the pump from a treatise on hydraulics and mechanics by a German Jesuit, Gaspar Schott, who described von Guericke's work only to refute it; Schott, a good Aristotelian like all 17th-century Jesuits, refused to contemplate the existence of a vacuum.

Boyle's imagination was caught by the account; he recognized the potential uses of an air pump more manageable than von Guericke's, the operation of which required the prolonged efforts of two strong men. He set his assistants the problem of constructing a laboratory version, and the result was the first deliberately designed air pump. Designed by Hooke, it was easily worked by one man and was moderately airtight. Its receiver, seven to eight gallons in volume, was made of glass and fitted so that objects could be readily put into it before pumping and then could be manipulated in the vacuum.

For more than 30 years natural philosophers had been trying to understand the working of suction pumps and siphons. For 25 years they had been demonstrating the effects of the pressure of

the air. Torricelli had performed his experiment in 1644. In it he had taken a tube closed at the bottom, filled it with mercury and then inverted it in a bowl of mercury; when the mercury moved partway down the tube, it was evident that the closed space above it could contain no air. Nearly 10 years earlier in Rome, however, the same experiment had been done with water. After Torricelli had described his experiment, the effect of atmospheric pressure was investigated often, and Pascal's Puy-de-Dôme experiment comparing mercury readings at different altitudes was repeated under varying conditions. French workers, notably the astronomer Adrien Auzout and the anatomist Jean Pecquet, had tried to use the Torricellian vacuum —the space above the column of mercury—for various experiments, but it was a cumbersome device at best. Boyle's vacuum now made possible a vast number of experiments. He could show how a deflated bladder swelled, how the mercury in the Torricellian tube fell and how the ticking of a watch suspended by a thread grew fainter and stopped as air was removed; how a burning candle went out and how a bird or kitten without air languished and eventually died. It seemed to Boyle that there was no end to the experiments that could be performed with his air pump—experiments,

moreover, that confirmed a "mechanical" view of the physical nature of air and supported the mechanical philosophy in general.

Boyle wrote up his experiments on air in 1659, and his report was published in 1660 under the title *New Experiments Physico-Mechanicall Touching the Spring of the Air, and its Effects, (Made for the most part, in a New Pneumatical Engine) Written by way of a Letter to the Right Honorable Charles Lord Viscount Dungarvan, Eldest Son to the Earl of Corke*. It was at one and the same time an admirable example of the kind of work English scientists had been doing in the 1650's and an appeal to the upper classes to take an interest in, and perhaps support, science. Boyle showed that the new experimental science could be gentlemanly, easy to read about and nontechnical, and at the same time precise and, to use the 17th-century expression, "fruitful." His was a new kind of scientific report; no one before Boyle had so carefully described the details of an experiment—the equipment, the specific manipulations and the precautions required to repeat it; no one before Boyle had so honestly confessed difficulties and errors. Although one did not need to be a scientist to repeat what he had done, few scientists could hope to do better; although many scientists rushed to have

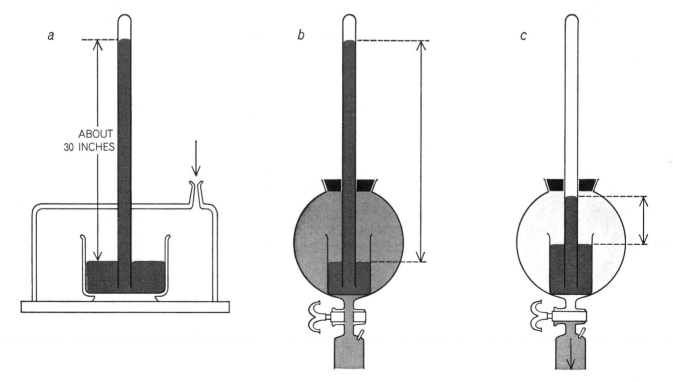

TORRICELLI'S EXPERIMENT showed that the pressure of the atmosphere would sustain about 30 inches of mercury (*a*). It was pressure, not the weight of the air, since stopping up the opening (*arrow*) made no difference. One of Boyle's objectives in devising

an air pump was to reduce the pressure of air on the exposed mercury surface. With a Torricellian tube placed inside the sealed receiver of his apparatus (*b*), he removed air from the receiver and reported that the level of the mercury column fell (*c*).

FIRST AIR PUMP, invented in 1650 by Otto von Guericke, was a suction pump that emptied a wine barrel filled with water (*top*). It did not really work. The second version removed air from a copper globe (*bottom*). In both cases operation required the sustained effort of two strong men. The engravings are from a book von Guericke published in 1672.

air pumps made for their own use, not even the great Huygens could go much beyond Boyle's original experiments with what came to be universally known as "the Boyleian vacuum."

The most others could do was to follow up Boyle's suggestion that from the effect of the vacuum on the Torricellian tube he might be able "to give a nearer guess at the proportion of force betwixt the pressure of the air and the gravity of the quicksilver," the gravity being "defined" in relation to the volume of air removed at each turn of the pump. Richard Towneley, a gentleman instrument-maker in the north of England who had been studying the Torricellian experiment, worked out the relation between the volume and pressure of rarefied air sometime after 1660. So did both Hooke and Lord Brouncker, the first president of the Royal Society. In the course of further demonstrating that the elasticity of the air could more than support the

weight of mercury required in a Torricellian experiment, Boyle performed his famous J-tube experiment, in which—compressing air rather than rarefying it—he established the validity of "the hypothesis, that supposes the pressures and expansions to be in reciprocal proportion" [*see illustration on opposite page*]. The account of his experiments, with "A table of the condensation of the air" and with due acknowledgment to the independent work of others, was published in 1662. What Boyle called his "hypothesis" is of course what we call Boyle's law: with temperature held constant, the volume of a gas is inversely proportional to the pressure on it. The hypothesis seems to have been first called Boyle's law by the Swiss mathematician Jakob Bernoulli in 1683, when the term "law" was just coming into scientific use.

When Boyle published further pneumatic experiments in 1669, he remarked that others seemed unable to construct

workable air pumps or perform significant experiments. Having given his original pump to the Royal Society, Boyle had done his new experiments with an improved second version. This model, in which some of the working parts were placed under water, maintained a better vacuum than the first; Boyle also attached various gauges with which to measure the degree of vacuum attained. Eleven years later he published still another work on pneumatics, this one particularly concerned with the effects of compressed air and the production of "factitious airs," such as carbon dioxide obtained from fermentation.

Boyle was a prolific writer in theology as well as science. His first book, *Seraphic Love* (1659), was much admired but is a juvenile and turgid work of devotion. Far more interesting today are his attempts to use science as a support for religion, as in *The Excellency of Theology compar'd with Natural Philosophy* (1674)—to which, characteristically, is annexed *Some Occasional Thoughts about the Excellency and Grounds of the Mechanical Philosophy*. Boyle's numerous other theological works all served science, if less directly, in that they helped to show that science did not lead to atheism and that opposition to Aristotle did not mean opposition to established religion. It is difficult now to see how any group with as many clergymen in it as the Royal Society could have been suspected of irreligion, but the society was nonetheless denounced from the pulpit, and its Fellows came to be touchy about any accusation of godlessness.

Perhaps stimulated by the establishment of the Royal Society and the fame of authorship, Boyle began in 1660 to gather up and publish his notes from earlier investigations. He maintained the informal style of the book on pneumatics and the practice of including a wealth of experimentation, carefully described. In all his works he sought to advance the mechanical philosophy by means of his old passion for chemistry. Indeed, the great aim of his life came to be the proof of the mechanical philosophy through chemical experiment and the creation of a science of chemistry through the introduction into chemical thought of the mechanical philosophy. For example, in *Certain Physiological Essays* (1661) Boyle begins with two essays on experimental science in general, continues with "Some specimens of an attempt to make Chymical Experiments Usefull to Illustrate the Notions of the Corpuscular Philosophy" and concludes with essays on the nature of fluidity and of firmness.

Here he demonstrates the reality of corpuscular entities by decomposing niter, or saltpeter (potassium nitrate, KNO_3), with a glowing coal and later reconstituting it. This was propaganda of a high order for the mechanical philosophy and a useful demonstration to nonchemists of the way chemical experimentation could be used to establish the truths of the mechanical philosophy. Indeed, it was the first time a chemical experiment had been used to prove the validity of a scientific theory!

In *The Sceptical Chymist*, also published in 1661, Boyle had two aims: to demonstrate the unreality of contemporary notions of elements (substances supposed to be found in all bodies and into which all bodies could be decomposed) and to introduce chemists to the virtues of the mechanical philosophy. There is nothing modern about Boyle's definition of an element, nor did he mean to define elements in any way that would be incompatible with 17th-century theory. What he wanted to show was that those things taken to be elementary (that is, the universal components of all substances, such as earth, air, fire and water or salt, sulfur and mercury) either were not stable or were not found in all substances, and that the only things that really conformed to the accepted definition of an element were fundamental particles. Chemists were only partly convinced, but they were influenced by

Boyle's novel experiments and by his manifest chemical competence.

Chemistry and the mechanical philosophy continued to go hand in hand for Boyle. In 1664 he published one of his most important works, *Experiments and Considerations Touching Colors*. He began with an exploration of the phenomena of whiteness and blackness, which he correctly declared to be the results of total reflection and total absorption of light. He demonstrated temperature effects with what was to become a classic series of experiments, including putting black and white cloths on snow and painting an object half black and half white and observing the temperature difference. When he attempted to discuss color in mechanical terms and show how the particulate structure of matter could be made to explain colors, he went partly astray; it was left to Hooke and later Newton to perceive that it was the light, rather than the colored object, that required analysis. Boyle studied the prismatic colors and tried reflecting and mixing them; it was left to Newton to refract them. So too he noticed the colors on soap bubbles and glass surfaces (Newton's rings), which first Hooke and later Newton undertook to explain in terms of light.

Because he focused his attention on matter rather than on light, Boyle was particularly interested in colored solutions; although he was therefore unable to make any advances in optics, he did make important contributions to chemistry. Studying changes in color achieved by mixing chemicals, he perceived for the first time that it was not some few particular acids that turned the blue of "syrup of violets" red but all acids, and only acids; all alkalis turned syrup of violets green. Then he found that all blue vegetable substances were similarly affected. He quickly saw that this provided not only an experimental basis for discussion of the mechanical interpretation of color but also a useful chemical discovery: if all acids turned blue vegetable solutions red, what did not do so was not an acid. And so Boyle proceeded to establish for the first time in chemistry a classification based on irrefutable chemical evidence: the division of substances into acid, alkali and neutral. (To recognize that some substances were neither acid nor alkali was a subtlety beyond many of Boyle's chemical contemporaries.) He went on to devise tests for distinguishing two of the common alkalis, potassium carbonate and ammonia, by another color change: the production of an orange or a white precipitate with a solution of mercuric chloride.

Boyle was soon in a position to work out methods of chemical analysis by combining these chemical tests with others, such as the blue imparted to solutions of copper, the characteristic shapes

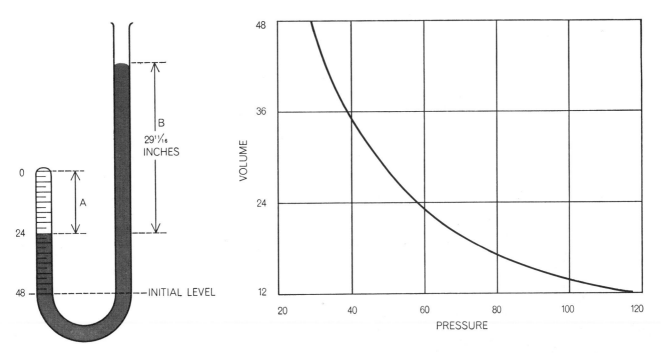

J-TUBE EXPERIMENT led to Boyle's law. Boyle noted the initial volume (48 units) of the air above the mercury on the closed side of the glass tube. Adding mercury through the open end, he saw that when the volume of the trapped air was cut in half (*A*), the mercury in the long arm of the tube was $29^{11}\!/_{16}$ inches higher (*B*) than in the closed side. From a table of such measurements (which yields the curve shown at the right) Boyle concluded that "the pressures and expansions" are "in reciprocal proportion," or that the volume of a gas is inversely proportional to the pressure on it ($P_1V_1 = P_2V_2$), provided that the temperature is kept constant.

of crystals, the specific gravity of fluids, the formation by silver of its chloride with "spirit of salt" (hydrochloric acid), the blackening of this compound on standing and so on. How such tests could be used he demonstrated in three later treatises: *Short Memoirs for the Natural Experimental History of Mineral Waters* (1685), a collection of tests and questions to be applied to unknown samples from new mineral springs, and *The Aerial Noctiluca* and *The Icy Noctiluca* (1680 and 1681), two wonderful experimental investigations of the element phosphorus. This mysterious substance had been discovered in Germany. Boyle saw some samples; being told that it derived from "somewhat to do with man's body," he deduced that the starting point must be human urine and soon produced a small quantity of the luminescent substance. His investigation was thorough and well organized, and he detected nearly all the properties of phosphorus that were to be known until the late 19th century: its flammability and how to control it by keeping the solid under water; its white and red forms; its solubility in various compounds, including certain aromatic oils; its ability to glow at low dilutions; its formation of phosphoric acid. *The Icy Noctiluca* deserves to be read as a classic example of early chemical analysis.

Besides such contributions to practical chemistry, which were much appreciated by his younger contemporaries, Boyle made important contributions to theoretical chemistry. His refusal to accept the concept of elements forced him into a totally new pattern of thought in which complex chemical substances were considered to be composed of simpler, known chemical substances rather than of mysterious, unknowable and vaguely defined elements. His analysis of chemicals in terms of "simple substances" such as the common salts, acids, metals and so on transformed chemistry, because analysis in these terms added to knowledge in a way that analysis in terms of hypothetical elements did not. The change in approach is illustrated by the fact that the early chemists of the French Academy of Sciences had spent two decades analyzing plants in the older style, inevitably reaching no conclusion other than that plants were all very similar in composition; Boyle's methods led French chemists in a similar stretch of time to reach the modern definition of a salt—a conclusion they might have reached earlier had they not at first regarded his turn of thought as too rationally physical to be serviceable in chemistry.

In 1668 Boyle had left the comparative quiet of Oxford for the busy life of London. In 1680 he was elected president of the Royal Society, but he turned down the office because of a scruple about taking an oath. Even without the title, however, he fulfilled the role of distinguished representative of English science. Learned foreigners came to visit; young scientists came to work in his laboratory. Through these young men—German, French and Dutch as well as English—his methods spread slowly but surely. His concept of a chemistry based on simple substances of definite composition that could be detected by experiment became universally accepted in the 18th century. He was, as a contemporary put it, "a mighty chemist."

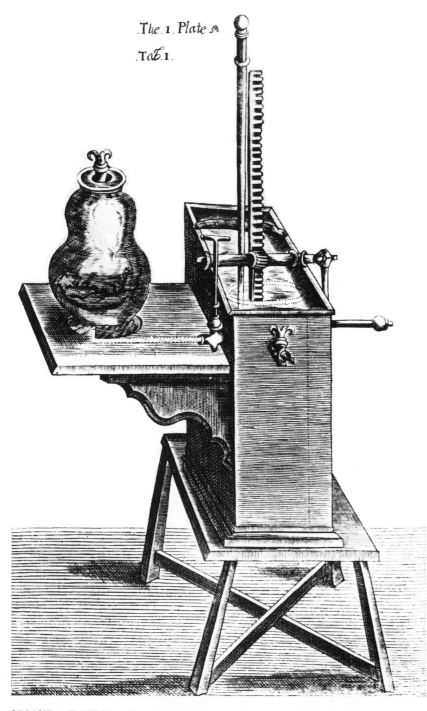

The 1. Plate

Tab 1.

SECOND AIR PUMP made for Boyle was somewhat more airtight than the first. The glass receiver (in which a kitten is seen in the engraving) stood separate, connected by tubing (*broken lines*) to the pump proper, the working parts of which were placed under water.

Lavoisier

by Denis I. Duveen
May 1956

This 18th-century Frenchman is best known as the founder of modern chemistry. He was also a remarkable biologist, farmer, technologist, financier, economist and politician

Antoine Laurent Lavoisier is universally known as the founder of modern chemistry, but this achievement tells only a small part of the story of his life. Had Lavoisier never performed a chemical experiment, he would still deserve a prominent place in history. He was a many-sided genius who pioneered not only in chemistry but also in physiology, scientific agriculture and technology, and at the same time was a leading figure of his era in finance, economics, public education and government. Few men in history have busied themselves in so many fields with such powerful effect as this brilliant and charming Frenchman.

Lavoisier was born in Paris on August 26, 1743, the only son of well-to-do parents. His mother died while he was still young, and he was brought up with loving care by his father and a maiden aunt. His father wanted him to become a lawyer; Antoine dutifully completed his legal education and obtained a license to practice, but he had shown his predilection for science by choosing to do his undergraduate work at the Collège Mazarin, where he studied astronomy, botany, chemistry and geology with famous masters. After law school he quickly turned to science again. Within three years, at the age of 25, he was elected to the Royal Academy of Sciences, as a result of his work in helping to prepare a geological atlas of France, his chemical research on plaster of Paris and his recognition with a special gold medal for plans submitted in a royal

ENGRAVING OF LAVOISIER was based on a painting made in 1788 by Jacques Louis David. On the ledge below the oval frame is some of Lavoisier's chemical apparatus.

competition to improve the street lighting of Paris.

Now resolved on a career of scientific research, Lavoisier first arranged to assure himself of sufficient financial means. He bought a share of the *Ferme Générale,* the private company that collected taxes for the King. This association was highly profitable to Lavoisier throughout his life, but it was to bring him to the guillotine.

At 28 Lavoisier married Marie Anne Pierrette Paulze, the 14-year-old daughter of a leading member of the *Ferme.* Although it was a marriage of convenience, arranged by the father to save his daughter from pressure in high places to marry an elderly and dissolute count, the union between Lavoisier and his child bride was to prove a happy success. Marie set about learning Latin and English to translate scientific works for her husband, who had little knowledge of foreign languages. She translated two important books by the Irish chemist Richard Kirwan and supplied Lavoisier with résumés of papers published by Joseph Priestley, Henry Cavendish and other contemporary British chemists. Her translations and footnotes show that she herself achieved more than a superficial knowledge of chemistry. As a hostess Marie made the Lavoisier home a popular meeting place for French and foreign scientists; as an accomplished artist she sketched and engraved plates for his books; she helped him in the laboratory as his secretary, taking notes on many of his experiments. After Lavoisier's execution she edited and printed for private circulation his last, uncompleted work, compiled in prison, *Mémoires de Chimie.* It seems a poor reward that her life after Lavoisier was made bitter by an unhappy, short marriage to Count Rumford, who was a re-

nowned scientist and inventor but also a careerist and adventurer.

Lavoisier's work in chemistry is a textbook classic which can be quickly reviewed. In 1772, at the age of 29, he began to study combustion and the "calcination" (oxidation) of metals. He observed that sulfur and phosphorus gained weight when they burned, and he supposed that they absorbed air. The key to explanation of his own observations came when Joseph Priestley discovered "dephlogisticated air" (oxygen). Lavoisier soon showed that it was this substance, to which he gave the name oxygen, that was absorbed by metals when they formed "calces," *i.e.,* oxides. He proceeded to replace the century-old "phlogiston" theory (that substances burned because of an escape of phlogiston) with the correct view that combustion is a chemical combination of the combustible substance with oxygen. Lavoisier could not explain the production of fire, and he introduced the term *calorique* to describe the *element imponderable*—heat. The complete explanation of combustion and heat was not to come until the theory of entropy was developed in the 19th century. Nonetheless Lavoisier, in collaboration with the great physicist Pierre Simon de Laplace, made studies of the heat evolved in combustion which laid the foundation of thermochemistry.

Lavoisier's theory at first failed to account for the combustion of "inflammable air" (hydrogen), evolved when metals were dissolved in acids. Here it was a discovery by Cavendish that gave Lavoisier the clue he needed. Cavendish learned that the burning of inflammable air produced pure water. Lavoisier extended his experiments and concluded that water was a compound of

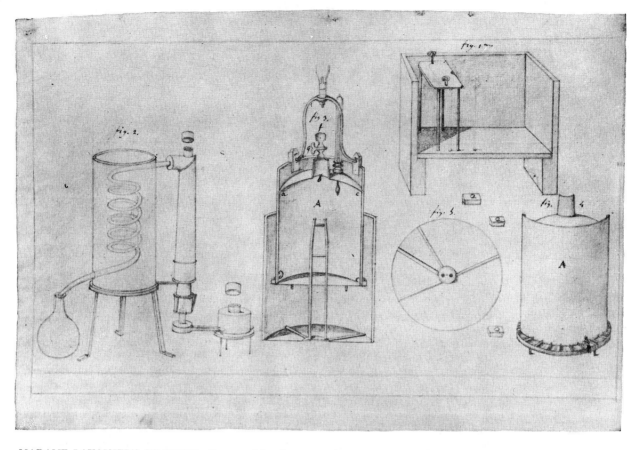

MADAME LAVOISIER'S DRAWING of some of her husband's chemical apparatus is shown above. At left is a device for condensing and collecting water formed during the combustion of alcohol. In the center is a gasometer. At upper right is a pneumatic trough. Shown below is a proof of the engraving made from the drawing. The objects are reversed by printing. The proof is corrected in Madame Lavoisier's hand. At the lower right it bears her signature: Paulze Lavoisier *sculpsit*.

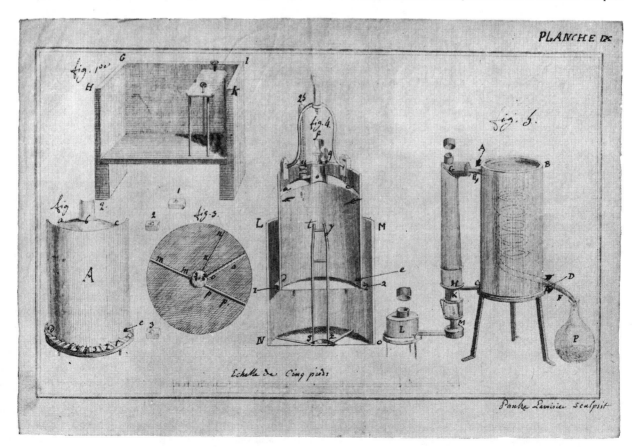

the two gases we now call oxygen and hydrogen. He recognized immediately that this fact supplied a keystone for the building of a whole new edifice of chemistry.

The new chemistry was quite readily accepted. It called for revision of the list of elements and a new system for naming substances; Lavoisier, with other leading French chemists, composed a new terminology, and with minor revisions it is still used today.

Lavoisier's keen interest in combustion led him naturally to respiration. There are those who say that his work in this field justifies his being called the founder of physiology and biochemistry. Certainly he brought order out of chaos. Many had guessed that all life depended on a vital ingredient in the atmosphere; Priestley and others had demonstrated by experiment that breathing animals exhausted the air of a necessary factor. It was left to Lavoisier to show the purely chemical nature of the role played by oxygen, or as it was first called, vital air, in respiration and combustion. He was the first to show that animal heat is produced by a slow

process continually occurring in the body and consisting of a form of slow combustion. To demonstrate this experimentally, he planned and carried out with Laplace a series of elegant experiments. They worked with guinea pigs. By accurately measuring the animals' intake of oxygen and output of carbon dioxide and heat—the latter with an ice calorimeter they invented—they laid the foundation of the science of calorimetry. As an extension of this work Lavoisier later collaborated with Armand Seguin in a program of research which established the facts of basal metabolism. The apparatus he designed for this work is the direct ancestor of the equipment used today in determining basal metabolism.

Lavoisier's scientific research was frequently interrupted by calls for technical assistance from the government. One of these was to remedy a shortage of gunpowder. France was suffering from a scarcity of saltpeter, an essential constituent of gunpowder, which was produced by an inefficient licensed monopoly. Called upon for advice by the

comptroller general of finance, Lavoisier suggested the formation of a government-owned *Régie des Poudres*. He was appointed as one of the four administrators of this agency and proceeded to institute new and efficient methods of production. Within three years he raised France's annual production of gunpowder from 714 tons to 1,686 tons. It can be said that Lavoisier's efforts contributed to the success of the American Revolution, for without the gunpowder supplied the colonists by France the outcome might have been different.

The *Régie des Poudres* provided Lavoisier with a home and a scientific laboratory at the Arsenal, where he spent his happiest and most productive years. But two episodes in this experience illustrate the hazards to which a scientist may be exposed in government service. On one occasion Lavoisier, his wife and three associates undertook to experiment with potassium chlorate as a possible new explosive. The experiments produced a laboratory explosion which killed two of the party, but the Lavoisiers escaped unharmed. Lavoisier reported the affair to the King's Minister

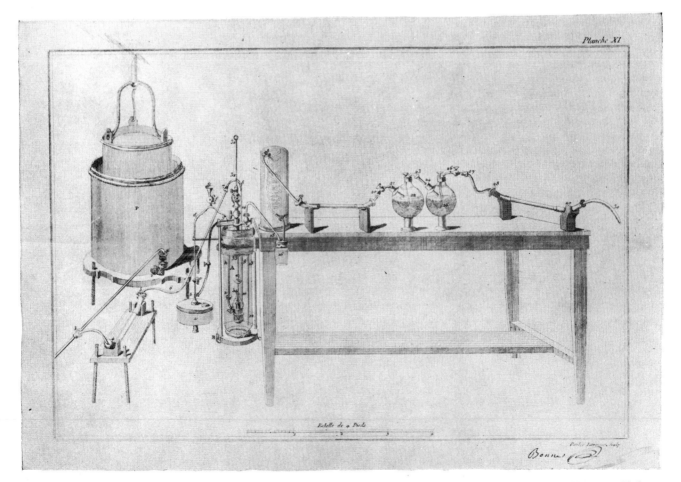

LAVOISIER'S APPARATUS for burning various oils in a measured quantity of oxygen and collecting the products of combustion are depicted in this plate from his *Elements of Chemistry*. Madame Lavoisier approved the plate by writing *Bonne* at lower right.

in lofty terms which well illustrate his character: "If you will deign, Sir, to engage the King's attention for a moment with an account of this sad accident and the dangers I faced, please take the opportunity to assure His Majesty that my life belongs to him and to the state, and that I shall always be ready to risk it whenever such action may be to his advantage, either by a resumption of the same work on the new explosive, work which I believe to be necessary, or in any other manner."

The second exposure was political. In 1789, when the Revolutionists had taken control of Paris, the Administration of Powders decided to ship 10,000 pounds of low-quality industrial gunpowder out of the city and replace it with better-quality musket powder. The move alarmed the populace; Lafayette, who was in charge of munitions and had not been informed of the shipment, ordered it returned to the Arsenal. The local commune investigated the powder administrators on charges of treason, and although the inquiry exonerated them, public clamor for Lavoisier's arrest did not abate until the powder was restored to the Arsenal.

Like Thomas Jefferson, whom he resembled in many ways, Lavoisier had a keen and personal interest in agriculture. He inherited from his father a farm at Le Bourget, and soon afterward he also acquired a large agricultural estate near the town of Orléans. Here he himself farmed about 370 acres and leased 865 acres to sharecroppers. It was his habit to spend the sowing and harvest seasons at the farm, and to keep close account of the crop yields and costs by double-entry bookkeeping. Farmer Lavoisier soon decided that crop yields were intimately connected with the amount of manure used on the fields. He then carefully calculated the optimum balance of cattle to acreage of pasture and cultivated land for a mixed farm. His studies of the requirements of various cash crops and of cattle were highly practical and successful. He was able to record with satisfaction that in 14 years he had doubled his yield of wheat and quintupled the size of his herd of cattle.

Lavoisier was active in the Agricultural Society of Paris and in the official Administration of Agriculture, of which he was one of the five original members and the guiding light. He represented the third estate in the Provincial Parliament of Orléans, where he was the prime mover of almost all the subjects discussed and decisions taken. His reports, which dominate the printed proceedings

REVOLUTIONARY PASSPORT was issued to Lavoisier in 1792 to enable him to travel to his farm near Blois. On the opposite side Lavoisier is described as 49 years old, five feet four inches, brown hair and eyes, long nose, small mouth, round chin, ordinary forehead and thin face. The words *le Roi* in the second line of the document have been crossed out.

of the Parliament, dealt not only with matters strictly agricultural but also with such varied subjects as public assistance for orphans and widows, steps to found a savings bank in Orléans, the abolition of the hated *corvée* (the obligation to repair the roads of a parish), tax reforms, the preparation of a mineralogical map of the district and the establishment of workhouses for the poor. He expressed his social creed in these words: "Happiness should not be limited to a small number of men; it belongs to all." Lavoisier was a physiocrat—devoted to the belief that all wealth stemmed from the land and that individual liberty was the most sacred right of man.

Scientifically a pioneer, politically a liberal, sociologically a reformer, Lavoisier was orthodox in his views on finance. In the new republic of 1789 he was elected to the presidency of the Discount Bank which was eventually to become the Bank of France. In a lucid and discerning address he noted with disquiet that inflation had set in. Three years later Lavoisier presented a report to the National Assembly on the lamentable state of the finances of the country. A recent appraisal of Lavoisier's exposition by an expert calls it superb. This report was printed by Lavoisier's friend Pierre S. Du Pont, whom he had financed in the publishing business and whose young son, Irénée, was an apprentice bookkeeper under Lavoisier at the Arsenal. When Irénée, after the Du Pont family's emigration to the U. S., established the great gunpowder works in Delaware, he wanted to name the factory Lavoisier Mills, but the family finally settled on E. I. Du Pont de Nemours.

Lavoisier's famous treatise on political economy, *On the Land Wealth of the*

Kingdom of France, is an important one in the history of economics. He had started it before the Revolution, but the National Assembly considered it so useful afterward that it ordered the paper printed in 1791. Lavoisier argued that a sound system of taxation could be founded only on exact knowledge of the country's agricultural production, and he collected data from all the provinces of France. His figures on production, consumption and population were the first reliable national statistics ever made available. Lavoisier recommended that France found an institution to gather and study all forms of economic data—not only on agriculture, but also on industry, population, capital and so on.

As a member of a committee established by the National Assembly in 1791 to advise the government on important questions concerning trades and crafts, Lavoisier proposed a national system of public education. He stressed that education of the people was a good investment from the state's point of view, and that free education should rightfully be available to all irrespective of sex and social position. He proposed the establishment of four kinds of schools: primary, elementary arts, institutes, and 12 national high schools, located in the 12 largest cities of France. He also suggested the creation of four national societies, to advance mathematics and the physical sciences, technical applications of science, the moral and political sciences, and literature and the fine arts.

Lavoisier took an active part in a little-known French attempt to establish an ambitious system of higher education in the new U. S. republic in 1788. The moving spirit of this project was Alexandre Marie Quesnay de Beaurepaire, grandson of a famous French philosopher, economist and court physician. Quesnay proposed a college, to be located in Richmond, the new capital of Virginia, which would be international in scope. The French Academy appointed a commission, which included Lavoisier, to study the question, and the commissioners made a favorable report, which it is reasonable to assume Lavoisier wrote, considering his propensity for taking the responsibility of drafting reports in all such situations.

Quesnay's academy was actually built in Richmond, but it never got a real start, probably because of the revolutionary overturn of France in the following year. It was in this very building that the Constitution of the U. S. was formally ratified. It later became a theater, burned down in 1811, was rebuilt, and today is in use as a church.

One of the first targets of the French Revolution—after the royal family—was the tax-collecting *Ferme Générale*, whose members had always borne odium as bloodsuckers who battened on the people. In 1791 the National Assembly finally suppressed the *Ferme* and ordered a detailed statement of its accounts. Delays in producing these accounts inflamed the Revolutionary Committee, and on November 14, 1793, arrest of all the Farmers-General was ordered. When Lavoisier heard of the decree, he went into hiding and tried to have the order reversed, on the grounds of his valuable scientific work for his country. But these attempts were fruitless, and after a few days he surrendered himself.

The Farmers-General were locked up in their former offices, where they finished the rendering of a final accounting in January, 1794. Their accounts showed quite clearly that the tax gatherers had acted throughout in complete conformity with the law.

The Terror, however, was entering upon its most extreme phase, and the Farmers-General were not to escape. New charges were preferred, accusing them of various abuses—levying excessive rates of interest, adulterating tobacco with excessive moisture (thus undermining smokers' health), and the like. In the heated atmosphere of the times the Farmers' accusers had no difficulty in getting a decree ordering their trial before the Revolutionary Tribunal. This was tantamount to a death sentence. At one o'clock in the morning of May 8, 1794, each of the prisoners was handed an almost illegible copy of the charges against him, and at 10 o'clock the same morning they were brought before the Tribunal for trial. Here a difficulty arose, for the Tribunal had jurisdiction only over counterrevolutionary activity, of which the Farmers-General had not been accused. But the Tribunal president, Jean Baptiste Coffinhal, disposed of the difficulty by charging the jury to ask themselves whether it had been shown that the Farmers had taken part in a plot against the people by various misdeeds, including supplying the enemies of the Republic with money illegally withheld from the Treasury—a charge which had not been mentioned in the indictment or supported by any evidence during the trial. The jury unanimously returned a verdict of guilty, and the convicted men were duly guillotined before nightfall.

So died France's greatest scientist. Joseph Louis Lagrange, the great mathematician, said the next day: "It required only a moment to sever that head, and perhaps a century will not be sufficient to produce another like it."

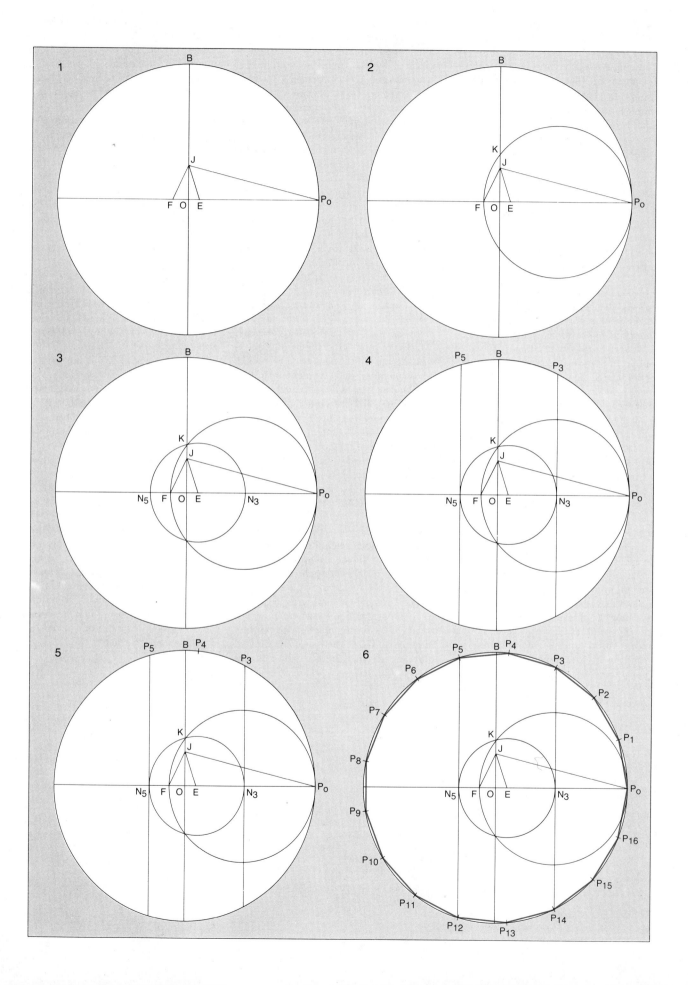

6

Gauss

by Ian Stewart
July 1977

*A child prodigy who became the leading mathematician
of his age, he was equally at home with the abstractions
of number theory, the long calculations of astronomy
and the practicalities of applied physics*

"Mathematics is the queen of the sciences," Carl Friedrich Gauss once said, and his own career exemplified that aphorism. Generally considered to rank with Archimedes and Newton as one of the ablest mathematicians of all time, Gauss was interested in both theory and application, and his contributions range from the purest number theory to practical problems of astronomy, magnetism and surveying. In all the branches of mathematics in which he worked he made profound discoveries, introduced new ideas and methods and laid the foundations of later investigations. It is a measure of his abilities that this year, the 200th anniversary of his birth, many of his ideas are still bearing fruit.

Gauss was in many ways an enigmatic and contradictory personality. The only son of working-class parents, he rose to become the leading mathematician of his age, yet he lived modestly and avoided public notice. His demeanor was mild, yet he was an aloof, politically reactionary and often unyielding man who asked only that he be allowed to continue his creative work undisturbed. He was prepared to recognize mathematical ability wherever he found it, in spite of contemporary prejudices, but his casual neglect of some of the best young mathematicians of his time, notably János Bolyai, one of the pioneers of non-Euclidean geometry, had unfortunate consequences.

A particularly striking aspect of Gauss's character was his refusal to present any of his work until he believed it had been polished to the point of perfection. No result, however important, was published until he deemed it to be complete. He reworked his mathematical proofs to such an extent that the path whereby he had obtained his results was all but obliterated. His published work has a quality of classical grace and elegance, austere and unapproachable. Many of his best ideas do not appear explicitly in print and have to be inferred by retracing the steps by which his discoveries must have been made. As a result important concepts did not see the light of day until they were discovered independently by others.

Gauss published some 155 titles during his lifetime, and he left behind a large quantity of unpublished work. In this brief survey I shall touch on his more important and influential discoveries and in some small measure attempt to explain how he arrived at them.

Gauss was born on April 30, 1777, in Brunswick (now in West Germany). His father worked variously as a gardener, a canal tender and a bricklayer. He was later described by his son as "an utterly honest and in many respects estimable and genuinely respectable man, but at home ... domineering, coarse and rude." Gauss's mother, the daughter of a stonecutter, was an intelligent woman of strong character. Her brother Friedrich played an important role in the life of the young Gauss. He worked as a weaver of fine damasks, but his interests were unusually broad. He spent much time encouraging Gauss and sharpening his wits in argument.

Among the great mathematicians there are about as many who showed mathematical talents in childhood as there are those who showed none at all until they were older. Gauss was unquestionably the most precocious of them all. He joked that he had been able to count before he could talk, and many anecdotes attest to his extraordinary gifts. It is said that one day before he was three years old his father was making out the weekly payroll for the laborers in his charge, unaware that his son was observing the process intently. Finishing his calculations, the elder Gauss was startled to hear a tiny voice saying: "Father, the reckoning is wrong, it should be...." When the computation was checked, the child's figure was found to be the right one. The remarkable thing about the story is that no one had taught him arithmetic.

Other tales concern Gauss's continued precocity at school. When he was 10, he was admitted to the arithmetic class. The master handed out a problem of the following type: What is the sum of $1 + 2 + 3 + ... + 100$, where there are 100 numbers and the difference between each number and the next is always one. There is a simple trick for doing such sums that was known to the master but not to his pupils.

The custom was for the first boy who solved a problem to lay his slate on the master's table with the answer written on it, for the next boy to lay his slate on top of that one and so on. The master had barely finished stating the problem when Gauss put his slate on the table. "There it lies," he said. For the next hour he sat with folded arms, getting an occasional skeptical look from the master, as the other boys toiled at the addition. At the end of the hour the master examined the slates. On Gauss's was a single number. Even in his old age Gauss loved to tell how of all the answers his was the only correct one.

To the master's credit he was so im-

GEOMETRIC CONSTRUCTION of a regular 17-sided polygon using only a straightedge and a compass, the first such discovery since Euclid, started Gauss on his mathematical career in 1796, at the age of 18. A simplified version of his construction, devised by H. W. Richmond in 1893, is shown in the illustration on the opposite page. The instructions are as follows: (1) Construct a circle with center at point O and radius OP_0 of arbitrary length. Draw OB perpendicular to OP_0. Find point J a quarter of the way up OB. Find point E such that the angle OJE is a quarter of the angle OJP_0. (This can be done by bisecting the angle twice.) Find point F such that the angle FJE is 45 degrees. (It is obtained by bisecting a right angle.) (2) Construct a circle with diameter FP_0. This circle intersects OB at point K. (3) Construct another circle with center at E and radius EK. This circle defines points N_5 and N_3. (4) Draw the lines N_3P_3 and N_5P_5 perpendicular to OP_0. (5) Bisect the arc P_3P_5 to obtain point P_4. (6) Using the chord P_4P_5, lay down that length sequentially starting at P_0. Connect points to obtain polygon.

pressed that he bought an arithmetic book with his own money and gave it to Gauss, who devoured it quickly. Gauss was also fortunate in that the mathematics master's assistant, a 17-year-old named Johann Martin Bartels, had a passion for mathematics, so that the two of them spent many hours studying together. On encountering the binomial theorem, which states that for all numbers n the expression $(1 + x)^n$ is a series, and that when n is not a positive integer, the series is infinite, Gauss was dissatisfied with the lack of rigor in the book the master had given him and constructed a proof. Although he was still a schoolboy, he was the first mathematician to pay serious attention to the problems of infinities. Few mathematical prodigies possess any ability beyond a facility for computation, but Gauss's abilities clearly extended into higher realms of thought.

At the age of 14 Gauss was introduced to the Duke of Brunswick, who had heard of his reputation and became his patron. The following year Gauss entered the Collegium Carolinum in Brunswick, where he studied and soon mastered the works of Newton, Leonhard Euler and Joseph Louis Lagrange. By the age of 19 he had discovered for himself and proved a remarkable theorem in number theory known as the law of quadratic reciprocity (to which I shall return). To appreciate how remarkable this was one must realize that although Euler had discovered the theorem earlier, both he and Adrien Marie Legendre had failed in their efforts to prove it.

When Gauss left the Collegium Carolinum in October, 1795, to study at the University of Göttingen, he was torn between mathematics and his other great love: the study of ancient languages, at which he was equally brilliant. On March 30, 1796, however, his mind was made up by one of the most surprising discoveries in the history of mathematics.

To provide some background for this discovery, let us go back two millenniums to Classical Greece. The main Greek contribution to mathematics was the flourishing school of geometry associated with the names of Pythagoras, Eudoxus, Euclid, Apollonius and Archimedes. The Greeks were probably the first to recognize the importance of rigor in proofs, and in search of such rigor they had imposed a number of restrictions. One of them was that in geometric constructions only a straightedge and a compass were to be used. In effect the only curves allowed were the straight line and the circle.

Euclid showed that it was possible to construct regular polygons with three, four, five and 15 sides, together with the polygons derived by repeated bisection of these sides, with a straightedge and a compass. Such polygons, however, were the only ones the Greeks could construct; they knew of no way to make polygons with, for example, seven, nine, 11, 13, 14 and 17 sides. For the next 2,000 years no one appears to have suspected that it might be possible to construct any of these other polygons. Gauss's achievement was to find a construction for a regular polygon of 17 sides, which he inscribed within a circle using only a straightedge and a compass. Moreover, he explained exactly which polygons could be so constructed: the number of sides must be any power of 2 (2^n) or a power of 2 multiplied by one or more different odd prime numbers of the type known as Fermat primes (after Pierre de Fermat, who discovered them). A prime number is by definition a number that cannot be divided evenly by any number except itself and 1; a Fermat prime has the additional requirement of being one greater than 2 to a power of 2, or $2^{2^n} + 1$. The only known Fermat primes are 3, 5, 17, 257 and 65,537. Thus we have the remarkable result that although regular 17-sided polygons can be constructed with a straightedge and a compass, regular polygons with seven, nine, 11, 13 and 14 sides cannot be.

Gauss proved this theorem (at the age of 18) by combining an algebraic argument with a geometric one. He showed that constructing a 17-sided polygon is tantamount to solving the equation $x^{16} + x^{15} + \ldots + x + 1 = 0$. Because 17 is prime and 16 is a power of 2 it turns out that this equation can be reduced to a series of quadratic equations (expressions of the form $ax^2 + bx + c = 0$, where a, b and c are given numbers and x is to be found). Since it had already been proved that quadratic equations can be solved with a straightedge-and-compass construction, the proof was

Λαμπάδια ἔχοντες διαδώσουσιν ἀλλήλοις.
PLATO.

C. F. Gauſs.   *Wilhelm Weber*

PORTRAITS of Gauss (*foreground*) and the physicist Wilhelm Weber (1804–91), with whom he collaborated on many practical experiments in magnetism and telegraphy, are shown in this elaborate etching. The Greek inscription on the ribbon reads: "God is an arithmetician." The Latin one at right reads: "The end crowns the work." The Greek quotation below, from the beginning of Plato's *Republic*, is translated: "Those who have torches will pass them to others."

complete. Apart from the proof's importance in inducing Gauss to take up mathematics as a career, it is the first real instance (beyond Descartes's introduction of coordinates) of a technique that has since become one of the most useful in mathematics: moving a problem from one domain (in this case geometry) to another (algebra) and solving it there.

By the time he was 20, Gauss once wrote, he was so overwhelmed with mathematical ideas that there was not enough time to record more than a small fraction of them. Many of those he managed to pursue appeared in his *Disquisitiones Arithmeticae*, published in 1801, when he was 24. This work may be said to have done for number theory what Euclid did for geometry: it organized scattered discoveries about integers and supplemented them with some deep ideas of Gauss's own. Gauss based his theory on the concept of congruent numbers, which are defined as two integers, *a* and *b*, that have the same remainder when they are divided by a given number *m*. Any two numbers that satisfy this requirement are said to be congruent "modulo" the number *m*, which is known as the modulus. For example, 16 and 23 both have a remainder of 2 when they are divided by 7, and they are therefore congruent modulo 7. Seven and 9 are congruent modulo 2, since there is a remainder of 1 when each number is divided by 2. Clearly any two even numbers and any two odd numbers will be congruent modulo 2.

Gauss also pointed out the possibility of doing arithmetic with congruent numbers. He showed that, for integers *a*, *b*, *c* and *d*, if, modulo a number *m*, *a* is congruent to *b* and *c* is congruent to *d*, then *a* + *c* is congruent to *b* + *d*, and *ac* is congruent to *bd*. Hence we can replace integers by congruent numbers in our arithmetic without running into contradictions. We do, however, get some surprises, such as the fact that, modulo 3, 1 + 1 + 1 is congruent to zero.

This arithmetic of congruent numbers is taught in many "new math" courses under the name modular arithmetic. I wonder how many teachers realize where it came from and why Gauss developed it? He wanted to use it as a tool to prove deep and difficult theorems. Perhaps the greatest of these, and certainly Gauss's personal favorite (he called it the golden theorem), is the law of quadratic reciprocity. Here some additional terminology is necessary.

Gauss first defined a "quadratic residue" by stating that if *m* is a positive integer and *a* is an integer that has no factors in common with *m*, then *a* is a quadratic residue of *m* if it is congruent, modulo *m*, to a perfect square. Another way of saying this is: If *a* is a quadratic residue of *m*, it is possible to find at least one *x* whose square divided by *m* leaves

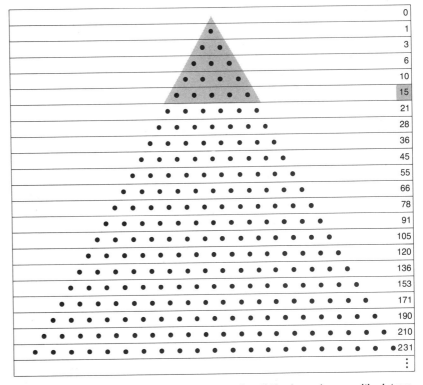

TRIANGULAR NUMBERS are numbers of form $n(n + 1)/2$, where n is any positive integer. They can also be represented as a triangular array of dots. In his *Disquisitiones Arithmeticae* Gauss proved that every positive integer is the sum of three such triangular numbers. He noted his discovery in his diary for July 10, 1796, with the cryptic entry: "Eureka! num $= \Delta + \Delta + \Delta$."

the remainder *a*. Thus 13 is a quadratic residue of 17 because the statement "x^2 is congruent to 13 modulo 17" is satisfied if *x* takes the value (among others) 8. Gauss proved that if *p* and *q* are different odd primes, then *p* is a quadratic residue of *q* if, and only if, *q* is a quadratic residue of *p*. There is a single exception to that rule: if both *p* and *q* are of the form $4n + 3$, then one is a quadratic residue and the other is not. This result may at first seem strangely specialized, but it enables us to decide whether a given odd prime is a quadratic residue of another prime by asking a question that in many cases is simpler: Is the second prime a quadratic residue of the first? This theorem has inspired some deep ideas of modern algebra and is of great importance throughout number theory and in other branches of mathematics. Gauss himself thought it so significant that during his lifetime he proved it eight different ways.

The *Disquisitiones* displayed a tendency that for Gauss became a way of life. Proofs were polished until they shone, with every trace of the process by which they had been reached removed, if it was at all possible, so that only the finished structure remained. Gauss once said: "When a fine building is finished, the scaffolding should no longer be visible." Later generations, faced with the problem of understanding Gauss's

methods as well as his results, may be forgiven if they accuse him of not only removing the scaffolding but also throwing away the plans. In mathematical investigation ideas and methods are often of greater importance than the theorems for which they were developed. A genuinely good idea can be extended into new fields to yield results that could not have been contemplated in advance. Here two aspects of mathematics are in conflict: mathematics as an art form and mathematics as a living discipline.

This view is not entirely a modern one. Gauss's contemporary Karl Gustav Jacobi said of him: "His proofs are stark and frozen...so that one must first thaw them out." Another contemporary, Niels Henrik Abel, remarked: "He is like the fox, who erases his tracks in the sand with his tail."

Why did Gauss conceal his methods, preferring to give only a synthesis and suppressing the analysis? He adopted the motto *Pauca sed matura* (Few but ripe), which reflects his dissatisfaction with the incomplete theorems of his colleagues. One cannot be sure, but perhaps the near-poverty of his childhood led him to be parsimonious with the distribution of his ideas. Perhaps too he did not want to display half-finished work for fear of being ridiculed if it should turn out to be wrong. This fear is common among the great mathematicians;

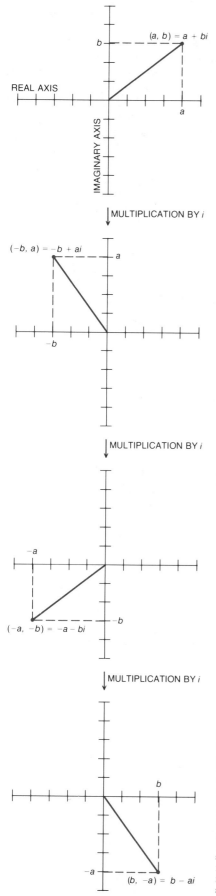

REAL AXIS

IMAGINARY AXIS

$(a, b) = a + bi$

MULTIPLICATION BY i

$(-b, a) = -b + ai$

MULTIPLICATION BY i

$(-a, -b) = -a - bi$

MULTIPLICATION BY i

$(b, -a) = b - ai$

Newton, for example, had to be persuaded to publish his *Principia*. It may even be justified; witness Georg Cantor, whose epochal work on set theory and transfinite numbers was ridiculed by leading mathematicians of his time. The experience seems to have led to his mental breakdown.

Much of Gauss's best work on number theory was related to the problem of complex numbers, which are defined as numbers of the form $a + bi$, where a and b are integers and i is the square root of -1 (so that $i^2 = -1$). Complex numbers were introduced by the Renaissance algebraists, but they were liberally endowed with mystical properties and issue-begging descriptions such as "real" and "imaginary." Even a man as intelligent as Leibniz was terribly confused by the subject, producing reams of nonsense about these mysterious numbers that are neither positive nor negative. "The Divine Spirit," he wrote, "found a sublime outlet in that wonder of analysis, that portent of the ideal world, that amphibian between being and not-being, which we call the imaginary root of negative unity."

Gauss was more prosaic, preferring to use a geometric representation of complex numbers as points in a plane. Although this approach had first been published in 1797, when a Norwegian surveyor, Caspar Wessel, devised an analytic representation of the geometry of the plane that was tantamount to complex numbers, his achievement was not noticed until 1897. A Swiss bookkeeper, Jean Robert Argand, developed a similar description in 1806, and he receives the credit in textbooks to this day. Gauss's contribution was to go beyond a purely geometric definition of complex numbers. In a letter he wrote in 1837 he says that in 1831 he had realized that a geometric interpretation can be avoided by using couples of the form (a, b) in place of $a + bi$, with purely algebraic definitions of the sum and the product. Using number pairs, he showed that the operations of arithmetic with complex numbers are defined by the rules: $(a, b) + (c, d)$ equals $(a + c, b + d)$ and $(a, b)(c, d)$ equals $(ac - bd, ad + bc)$. It is

COMPLEX NUMBERS of the form $a + bi$ (where a and b are integers and i is the square root of -1) can be represented as number pairs (a, b), or points of a plane, in the same way that real numbers are represented as points of a line. The complex number pairs can then be manipulated in a geometric fashion. For example, the rotation through 90 degrees of a line joining the origin and the point (a, b) is equivalent to multiplying the original complex number by i. (Three such rotations are shown in the illustration.) Gauss was the first mathematician to notice that this geometric interpretation could be used to obtain purely algebraic definitions of the sum and product.

easy to verify that the couple $(a, 0)$ behaves just like the real number a, and that $(0, 1)^2$ equals $(-1, 0)$. Thus the number pair $(0, 1)$ is our mysterious square root of -1. The true value of this approach, which is the one generally adopted today, was not appreciated until William Rowan Hamilton published it in 1837. Gauss was the first mathematician to make free and extensive use of complex numbers and to give them full acceptance as genuine mathematical concepts.

In his doctoral dissertation for the University of Helmstedt, awarded in 1799, Gauss gave the first proof of the "fundamental theorem of algebra" (today more naturally proved as a theorem in topology): that every polynomial equation has a complex root. Here again Gauss thought the theorem was so important that during his lifetime he proved it four different ways. The third proof is particularly characteristic of his impenetrable style and original turn of mind. He constructs from a polynomial equation a complicated expression in the form of a double integral. If the polynomial has no root, this integral should take the same value whether it is integrated with respect to first one variable and then the second, or the other way around. Gauss shows that this is not the case, that the two processes give different results; therefore the assumption of no root is false, and hence a root exists.

To see where the proof came from we have to know that Gauss possessed the basic theorems of complex-number analysis but had not published them. His proof is a translation of an argument in complex-function theory into the more standard terms of real-number analysis. The results of this translation are logically rigorous, but they involve a certain perversity of formulation that obscures the motivating principles. Apparently Gauss felt that his theorems of complex-number analysis were not sufficiently complete for publication, and he recast his arguments accordingly.

Gauss did much more with complex numbers. In 1811 he discovered what is now called Cauchy's theorem: The integral of a complex analytical function around a closed curve that encloses no singularities is zero. Augustin Louis Cauchy made it the foundation of complex-number analysis, and it still is. We do not call it Gauss's theorem because Gauss never published it. It seems likely that he intended to produce a definitive work on complex-number analysis but never found the time to work out the ideas to his own satisfaction.

Gauss also developed a method of factoring primes with complex numbers, with some intriguing results. The prime number 2, for example, can be factored in the form $(1 + i)(1 - i)$. In the same way, 5 can be factored as

$(2 + i)(2 - i)$, 29 as $(5 + 2i)(5 - 2i)$, and so on. Certain prime numbers, however (among them 7, 11 and 19), cannot be factored and remain prime. Gauss discovered that only primes of the form $4n + 1$ can be factored with complex numbers (the number 2 is a special case) and that each of these primes can be factored in exactly one way. Such methods were later used to solve problems that on the face of it did not involve complex numbers at all.

In particular Gauss used complex numbers of the form $a + bi$ (now called Gaussian integers) to formulate and prove a version of the law of quadratic reciprocity for biquadratic residues. The number k is said to be a biquadratic residue of another number m if k is congruent modulo m to the fourth power of an integer. Thus the biquadratic residues of 10 are 0, 1, 5 and 6. The law of biquadratic reciprocity states that for two odd prime numbers p and q there are connections between the statement "p is a biquadratic residue of q" and the statement "q is a biquadratic residue of p," with a host of conditions on the form of p and q. This theorem is analogous to the law of quadratic reciprocity, but it is much more cumbersome to state mathematically (and hence is very hard to conjecture, let alone to prove). If the theorem is extended to the case where p and q are Gaussian integers of the form $a + bi$, there are remarkable simplifications both in the statement of the result and in its proof. Thus going over to the case of complex numbers makes the problem easier and is more natural than employing the case of real numbers.

Gauss's proof of biquadratic reciprocity using Gaussian integers provided an archetypal procedure for solving many problems in number theory: First,

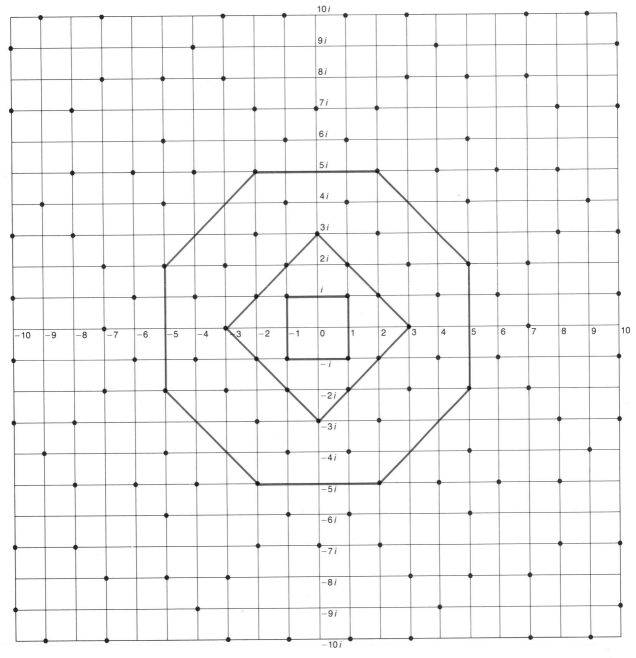

GAUSSIAN PRIMES, complex numbers of the form $a + bi$ having no factors of the same type, are distributed irregularly over the complex plane. Gauss discovered three classes: (1) $\pm 1 \pm i$, forming the vertexes of a square, (2) $\pm p$ and $\pm pi$ (where p is a real prime number equal to $4n + 3$), forming a diamond, and (3) $\pm a \pm bi$ and $\pm b \pm ai$, forming a truncated square. Gaussian primes always occur in one of these patterns.

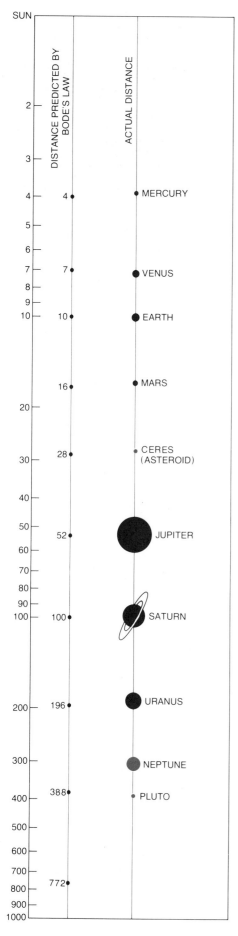

SUN

DISTANCE PREDICTED BY BODE'S LAW

ACTUAL DISTANCE

4 • MERCURY

7 • VENUS

10 • EARTH

16 • MARS

28 • CERES (ASTEROID)

52 • JUPITER

100 • SATURN

196 • URANUS

NEPTUNE

388 • PLUTO

772 •

DISTANCES OF THE PLANETS from the sun are approximated by Bode's law, originally discovered by Johann Titius in 1766. (One unit equals 9.3 million miles.) The gap in the series known in 1800 precipitated the search for the "missing planet," culminating in the discovery of the asteroid Ceres in 1801. Gauss carried out the difficult task of computing the orbit of Ceres from the scanty data available. It reappeared where he predicted. (Planets discovered after 1800 are in color. Note that Bode's law poorly approximates the locations of the planets beyond Uranus.)

extend the theorem to a domain of suitably chosen complex numbers called a number field, where it may be dealt with more naturally, solve it there and return to the case of ordinary integers at the end of the proof. This powerful method opened the door to what is now called algebraic number theory.

In 1801 Gauss's mathematical interests changed direction sharply when he became involved in astronomy. His love of calculation was clearly a contributory factor. Throughout his most erudite works are long calculations, and some of his deepest theorems in number theory were inferred by examining long lists of figures. Moreover, Gauss carried many of his calculations out to 21 decimals long before the availability of any kind of computer.

Gauss's interest in astronomy can be traced back to a discovery by Johann Titius, who in 1776 stated an empirical rule for the distances between the sun and the planets. Titius started with the series 0, 3, 6, 12, 24, 48 and 96, in which each term after the first is double the preceding one; then he added 4 to each term to get 4, 7, 10, 16, 28, 52 and 100. As it turned out, these numbers were very closely proportional to the distances from the sun of Mercury, Venus, the earth, Mars, Jupiter and Saturn, except that there was no planet at distance 28. This rule, which came to be known as Bode's law (because Johann Bode appropriated it without acknowledgment), remained a curiosity until 1781, when William Herschel discovered Uranus at a distance of approximately 196 units. Since the next term in the Titius-Bode sequence was $2(96) + 4 = 196$, interest now focused on the gap at distance 28.

On New Year's Eve, 1800–1801, Giuseppe Piazzi discovered what he thought was the missing planet. It was Ceres, which we now know as one of the thousands of small bodies in the asteroid belt between Mars and Jupiter. Once Ceres was sighted it became important to calculate the elliptical orbit of the new body before observers lost track of it. The difficulty of observing such a small object made the available data scanty and the calculation of the orbit all the harder; Newton himself had remarked how difficult it was to compute orbits

from meager data. It was a golden opportunity for Gauss to follow in the footsteps of the man he admired most.

After only three observations Gauss developed a technique for calculating orbital components with such accuracy that late in 1801 and early in 1802 several astronomers were able to locate Ceres again without difficulty. As part of his technique he showed how the variation inherent in experimentally derived information could be represented by a bell-shaped curve (best known today as the Gaussian distribution). He also developed the method of least squares, by which the best estimated value is derived from the minimum sums of squared differences in a particular computation. His methods, which he described in an 1809 paper titled "Theory of the motion of the heavenly bodies revolving around the sun in conic sections," are still in service today. Only a few modifications were required to adapt them to modern computers.

Gauss had similar success in determining the orbit of the asteroid Pallas, for which he refined his calculations to take into account the perturbations caused by the other planets in the solar system. In 1807 he became professor of astronomy and director of the new observatory at the University of Göttingen, where he remained for the rest of his life. His first wife died in 1809, soon after the birth of their third child. From a second marriage were born two sons and a daughter.

In about 1820 Gauss turned his attention to geodesy: the mathematical determination of the size and shape of the earth. To it he devoted much of the next eight years in theoretical studies and fieldwork. In 1821 he became scientific adviser to the governments of Hannover and Denmark, which charged him with making a geodetic survey of Hannover by means of the technique of triangulation. To this end he developed the heliotrope, a device that reflected the rays of the sun in a precisely specified direction, thus making it possible to accurately align surveying instruments over long distances.

Gauss's efforts to determine the shape of the earth by actual geodetic measurements led him back into pure theory. Working with data from his surveys, he developed a theory of curved surfaces in which the characteristics of a surface can be determined solely by measuring the length of the curves that lie on it. This intrinsic-surface theory inspired one of his students, Bernhard Riemann, to develop a general intrinsic geometry of spaces with three or more dimensions. Some 60 years later Riemann's ideas formed the mathematical basis for Albert Einstein's general theory of relativity.

Beginning in 1831, when the physicist Wilhelm Weber arrived in Göttingen, Gauss worked closely with him on ex-

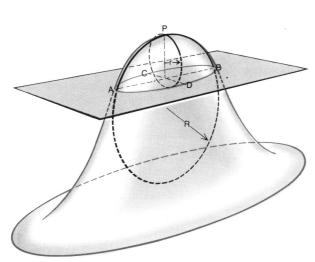

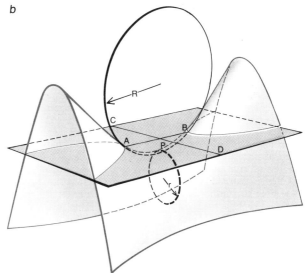

a

b

INTRINSIC SURFACE THEORY developed by Gauss enables one to compute the curvature of a surface solely by measuring the length of the curves that lie on that surface. The curvature of a convex surface (a) is found as follows: First, using a plane parallel to the tangent plane at point P, slice the surface near P along an ellipse (*dark color*). Draw the major and minor axes of the ellipse, designated AB and CD, and project these axes onto the surface to obtain the curves APB

and CPD. Next find two circles having the closest fit at point P to the projected curves. If the radii of the two circles are R and r respectively, then as the plane approaches P and the ellipse gets very small the curvature will approach $1/Rr$. In b the surface is saddle-shaped and the plane cuts it at two hyperbolas. By convention one radius (and hence the curvature) is negative. For ordinary flat space the two radii R and r are infinite in length and the curvature is zero.

perimental and theoretical investigations of magnetism. They invented a magnetometer, and as a result of their interest in the magnetism of the earth they organized a network of observers across Europe to measure variations in the terrestrial magnetic field. Gauss was

able to show by a theoretical analysis that the field arose within the solid earth, a result of considerable importance because it limited the field's possible origins and focused attention on geophysical mechanisms of its generation. His contribution is recognized in the gauss,

the unit of the density of magnetic flux.

Gauss and Weber were also among the first to point out the possibility of sending messages by electricity. The history of telegraphy is long, but before 1800 or so only nonelectric methods were in use. Then in 1827 an electric

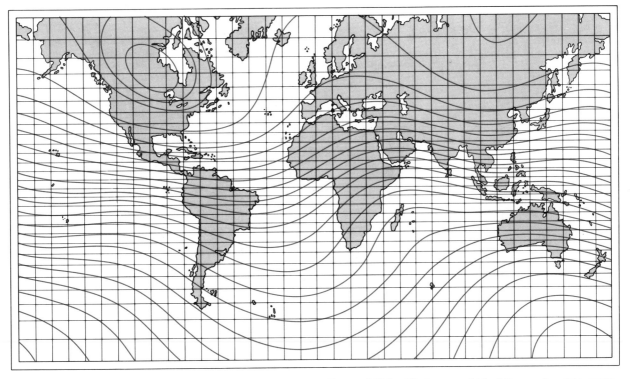

MAGNETIC-FIELD MAP of the earth is based on a drawing that appeared in 1840 in *Atlas des Erdmagnetismus*, written jointly by Gauss and Weber. The two men obtained the experimental measurements for the map by organizing a worldwide network of observers.

impulse was transmitted over a distance of a sixth of a mile, which immediately suggested that electricity could serve for telegraphy. Various electric telegraphs were designed, but none was implemented until 1832, when the Czar's summer and winter palaces in St. Petersburg were linked. A year later Gauss and Weber had a telegraph running over the rooftops of Göttingen for a distance of 2.3 kilometers. The signals were transmitted as a sequence of five deflections of a needle either to the right or to the left (a total of 32 possibilities), and the device worked so well that the two men used it routinely to communicate. The Gauss-Weber telegraph was probably the first to work in any useful sense, and it antedated Samuel F. B. Morse's famous patent by seven years.

Gauss's great reputation was still further enhanced after his death by the discovery of unpublished results anticipating many of the major advances of the 19th century. In addition to Cauchy's theorem he had discovered the double periodicity of elliptic functions, which in the hands of Jacobi and Abel became the center of 19th-century func-

tion theory. The elliptic functions are certain special functions $f(z)$ of a complex number z. Double periodicity means that there are two distinct complex constants, say a and b, such that for any z, $f(z)$ equals $f(z + a)$, which also equals $f(z + b)$. This is analogous to the single periodicity of the trigonometric functions, so that $\sin(z)$ has the property $\sin(z + 2\pi) = \sin(z)$. Gauss's discovery had wide implications for the connections between number theory and complex-function theory.

Gauss was also among the first to doubt that Euclidean geometry was inherent in nature and human thought. Euclid's systematic geometry had been based on certain axioms, or fundamental propositions, that were regarded as self-evident truths. On these unproved foundations the entire system was constructed through pure logic. The parallel axiom states that only one line can be drawn parallel to a given line through any point not on that line. Throughout history the parallel axiom was intensively investigated because, unlike the other axioms, it was not immediately persuasive and required the concept of infinity. Even in antiquity mathematicians tried

in vain to replace it with more obvious axioms.

In the 18th century there was a renewed interest in this unsolved problem, and many mathematicians and amateurs attempted to prove that the parallel axiom followed logically from Euclid's other axioms. All these proofs turned out to be fallacious. Gauss became aware of the controversy as a student at Göttingen, and in 1804 he wrote a letter to the Hungarian mathematician Farkas Bolyai, showing that Bolyai's proof of the parallel axiom was false because he had replaced an infinite argument by a finite one. Gauss accompanied his refutation with a comment to the effect that he had been troubled by the same difficulty. In 1815, however, he wrote certain book reviews in which he clearly implied that a geometry might exist where the parallel axiom did not hold, yet which would still be internally consistent and free of contradiction. Given Gauss's usual caution about revealing his ideas, this statement strongly suggests that he had some proof to back it up. Perhaps because his ideas ran counter to contemporary views he chose to suppress them; he may have felt, probably correctly, that they would be misunderstood.

In 1820 Bolyai's son János, who had been infected by his father's fanatical preoccupation with proving the parallel axiom, concluded that a proof was impossible and began to develop a new geometry that did not depend on Euclid's axiom. Three years later he completed a paper proposing a consistent system of non-Euclidean geometry, which was published as an appendix to his father's book *Essay on the Elements of Mathematics for Studious Youths*. Gauss read that paper in 1832 and wrote the elder Bolyai that he could not praise it because doing so would be tantamount to praising work he had done himself 30 years before. The younger Bolyai was profoundly disappointed by Gauss's rejection, and he died essentially unrecognized for his solution to an enormously important and long-standing problem (solved independently at about the same time and in much the same way by Nikolai Ivanovich Lobachevsky). Gauss's attitude was quite unfair in view of the fact that he had never felt confident enough in his own work to publicize it. Perhaps he was a bit jealous of Bolyai's success.

In many ways Gauss stood at a crossroads. He can be viewed equally well as either the first of the modern mathematicians or the last of the great classical ones. The paradox can easily be resolved: his methods were modern in spirit but his choice of problems was classical.

A trademark of Gauss's work, particularly in pure mathematics, is his reasoning with the particular as if it were the general. The success of this technique depends on using only those prop-

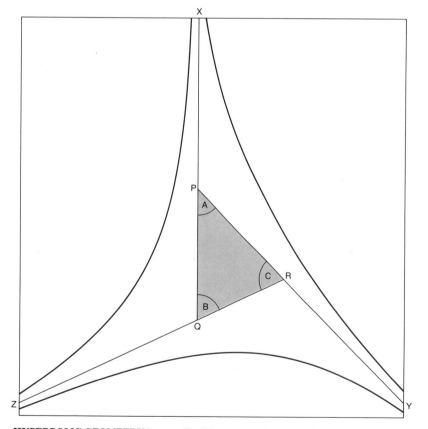

HYPERBOLIC GEOMETRY is a non-Euclidean system devised by Gauss. It enables one to find the area of a triangle from its angles, a result that fails in Euclidean geometry. The diagram shown here is part of Gauss's proof that the area of the triangle is proportional to the difference between 180 degrees and the sum of its angles *A*, *B* and *C*. Because the diagram is drawn on a Euclidean plane the lines appear curved, but in hyperbolic space they are "straight." Moreover, the lines occurring in triplets at the edges of the diagram are all parallel, so that the apparently curved line *XY* is parallel to both *XQ* and *PY*, a situation that cannot occur in Euclidean geometry. Although the triangle *XYZ* has all its vertexes at infinity, its area is finite.

erties of a special case that have general counterparts. Gauss's approach combines the breadth of general methods with the intensity and simplicity of particular examples. Thus his work on complex numbers has within it the seeds of a general theory of algebraic numbers. His writings always leave one with the feeling that he knows more than he is telling, that when he explains a result, he already has an idea of more general questions surrounding it and of how to make a start on solving them.

A measure of the depth and fertility of Gauss's ideas can be obtained from recent investigations inspired by them. For example, in 1947 André Weil, starting from some theorems of Gauss's on the number of solutions to polynomial equations modulo a prime number, was led to formulate three far-reaching conjectures about algebraic varieties over finite fields. A finite field is a set of algebraic elements, finite in number, that, in addition to meeting certain other requirements, can be added, multiplied, subtracted and divided to give quantities of the same type. Thus the integers modulo a prime number p form such a field with p elements.

The Weil conjectures give formulas for the number of solutions to an algebraic equation in a finite field. In particular they allow one to deduce that a given equation does or does not have solutions; this information can be transferred to analogous equations involving integers or algebraic numbers. Of course, the actual formulation of the Weil conjectures is very technical. They have recently been proved by Pierre Deligne, and they are of fundamental importance in algebraic geometry.

Elsewhere Gauss conjectured that prime factorization was unique for numbers of the form $p + q\sqrt{-D}$, where D is a positive integer, only when D is equal to one of the numbers 1, 2, 3, 7, 11, 19, 43, 67 and 163. This result, which he inferred from numerical evidence, has recently been proved by Harold M. Stark and Alan Baker and leads to important new theorems in number theory.

Gauss's importance stems from this mixture of the general and the specific. He forms a bridge between old and new, and his ideas contain within them the seeds of spacious theories and important results. He is the prime instance of a mathematician who could extract the maximum amount of juice from a ripe example by inductive reasoning (working from a particular case to a general principle) rather than deductive reasoning (drawing a specific conclusion from general principles). Of Gauss's influence on his successors Eric Temple Bell wrote in 1937: "He lives everywhere in mathematics." If anything, that is truer today than it was when Bell said it 40 years ago.

7

The Short Life of Évariste Galois

by Tony Rothman
April 1982

Legend has it that the young mathematician wrote down group theory the night before he was fatally shot in a duel. More careful investigation shows that Galois's remarkable ideas took somewhat longer to mature

In the early morning hours of May 30, 1832, the French mathematical prodigy Évariste Galois, who was then 20 years old, wrote to his friends Napoleon Lebon and V. Delauney:

"I have been provoked by two patriots.... It is impossible for me to refuse. I beg for your forgiveness for not having told you. But my adversaries have put me on my word of honor not to inform any patriot. Your task is simple: prove that I am fighting against my will, having exhausted all possible means of reconciliation; say whether I am capable of lying even in the most trivial matters. Please remember me since fate did not give me enough of a life to be remembered by my country.

I die your friend,
É. Galois"

On the same night Galois also wrote to his friend Auguste Chevalier:

"I have made some new discoveries in analysis. The first concerns the theory of equations, the others integral functions.

"In the theory of equations I have investigated the conditions for the solvability of equations by radicals; this has given me the occasion to deepen this theory and describe all the transformations possible on an equation even though it is not solvable by radicals. All of this will be found here in three memoirs....

"Make a public request of [Carl Gustav Jacob] Jacobi or [Carl Friedrich] Gauss to give their opinions not as to the truth but as to the importance of these theorems. After that, I hope some men will find it profitable to sort out this mess."

Galois's desperate state during the writing of these letters was fully warranted in view of the subsequent events. Shortly after sunrise on the morning he completed the letters he left his room at the pension Sieur Faultrier in Paris and confronted a political activist named Pescheux d'Herbinville in a duel of honor on the banks of a nearby pond. There

Galois was shot in the abdomen and abandoned. A passerby found him and he was taken to the Hôpital Cochin, where he died the next day. Fourteen years later the manuscripts he had left behind for Chevalier were published by the French mathematician Joseph Liouville, and the extraordinarily fecund branch of mathematics called group theory was born.

Few tales in the history of science can equal the high romance of the known facts about the life and death of Galois. Yet because the facts of the story are so compelling it is easy to read too much into Galois's letters, and it is tempting to sift through the events that led up to the duel for an explanatory thread that can match the melodrama apparent in his writings.

It is known, for example, that at age 17 Galois was instrumental in creating a branch of mathematics that now provides insights into such diverse areas as arithmetic, crystallography, particle physics and the attainable positions of Rubik's cube. It is also a matter of record that at the same age Galois failed for a second time the mathematics examination for admission to the École Polytechnique. He studied instead at the École Normale in Paris, but by the time he was 19 he had been expelled from the school and twice arrested and imprisoned for his political activities. Shortly before the duel he had become involved in an unhappy love affair, which in one of his last letters he seemed to link with the duel itself. "I die," he wrote, "the victim of an infamous coquette and her two dupes."

Unfortunately several of Galois's 20th-century biographers have not resisted the temptation to arrange, interpret and embellish such facts. The story of Galois known to most people today is derived from popular accounts, such as those by the physicist Leopold Infeld and the astronomer Fred Hoyle.

The most influential version of the story has been that of Eric Temple Bell, the mathematician whose 1937 book *Men of Mathematics* is probably the best-known introduction to the lives of great mathematicians.

In the popular retellings of the tale Galois is presented as a misunderstood genius, oppressed by the stupidity of his teachers, ignored by the mathematical establishment and goaded by the events of the times into political activities that squandered his energies and eventually cost him his life. Most remarkable of all, according to these accounts, is that throughout the political turmoil and even during his imprisonment Galois continued to develop his mathematical ideas in his head and finally wrote them down the night before the duel. Bell's description of the final night is worth quoting because it has probably given the greatest impetus to the Galois myth:

"All night long he had spent the fleeting hours feverishly dashing off his scientific last will and testament, writing against time to glean a few of the great things in his teeming mind before the death which he saw could overtake him. Time after time he broke off to scribble in the margin 'I have not time; I have not time,' and passed on to the next frantically scrawled outline. What he wrote in those last desperate hours before the dawn will keep generations of mathematicians busy for hundreds of years."

Recently, with the help of Marc Henneaux and Cecile DeWitt-Morrette of the University of Texas at Austin, I have studied some of Galois's writings and the later scholarly work on his life. Although it is clear from these materials that the major events in Galois's life have been known for some time, the reconstructions by Bell and others reveal more about the stereotypes of scientific genius that appeal to the popular imagination than they do about Galois. The true romance of Évariste Galois is a fascinating story in its own right, and it bears telling on the 150th anniversary of his death.

Apart from letters, official records and other contemporary documents the principal source on the life of Galois is an 1896 biography by Paul Dupuy, a historian and the general superintendent of the École Normale, the college Galois had attended 66 years earlier. According to Dupuy, Galois was born on October 25, 1811, in Bourg-la-Reine, a suburb of Paris. His father, Nicholas-Gabriel Galois, supported Napoleon and headed the town's liberal party; he was elected mayor of Bourg-la-Reine in 1815 during the Hundred Days, Napoleon's first return from exile.

For the first 12 years of his life Évariste was educated by his mother, Adeläide-Marie Demante Galois. She gave her son a solid background in Greek and Latin, and she passed on to him her skepticism toward established religion. It is unlikely, however, that the young Galois was exposed to mathematics in any more detail than the usual lessons in arithmetic; a mathematics education was not considered particularly important at the time. There is no record of previous mathematical talent on either side of the family.

Galois's formal education began in 1823, when he was enrolled in the Collège Royal de Louis-le-Grand, the Paris preparatory school that was the alma mater of Robespierre and Victor Hugo (and is still operating today). At Louis-le-Grand, Galois immediately began to develop his political sensibilities. His liberal, or antiroyalist, sympathies, acquired from his parents, were in accord with the political opinions of most of the other students.

During Galois's first term, however, relations between the students and a newly appointed headmaster of the school were badly strained. The students suspected the headmaster of planning to return the school to Jesuit administration; the Jesuits were leaders of the right-wing backlash that followed the Napoleonic era. The students staged a minor rebellion: they refused to sing at chapel, to recite in class or to toast Louis XVIII at a school banquet. The headmaster summarily expelled 40 students he suspected of leading the insurrection. Although Galois was not expelled (and it is not known whether he participated in the uprising), the arbitrary action of the headmaster undoubtedly helped to foster Galois's distrust of authority.

There is little evidence that Galois was a poor student or that his intellectual growth was stunted by inferior teachers at Louis-le-Grand, as the popular accounts would have it. In his first few years he won several prizes in Greek

DRAWING OF GALOIS by David A. Johnson depicts the mathematician at the age of 17, while he was a student at the Collège Royal de Louis-le-Grand. At the time Galois had studied mathematics for only two years, yet he had already published a paper on continued fractions and had begun the investigations into the theory of equations that led him to consider an abstract algebraic theory of sets of objects he called groups. Credit for the development of group theory must also be given to several other mathematicians of the late 18th and early 19th centuries, notably Paulo Ruffini, Neils Henrik Abel and Joseph Louis Lagrange. The distinction of founding group theory, however, is usually accorded to Galois. Johnson's drawing is based on the two known renderings of Galois. One was done when Galois was 15 and the other was completed from memory by Galois's brother Alfred in 1848, 16 years after Évariste's death.

and Latin and half a dozen honorable mentions. The historian of science René Taton calls his progress brilliant. Nevertheless, during Galois's third year his work in rhetoric was inadequate and he had to repeat the year. Contrary to Bell's statement that Galois's poor work in rhetoric was a result of his preoccupation with algebra, it was only after this setback that Galois enrolled in his first course in mathematics. He was then 15.

The course, taught by Hippolyte Jean Vernier, awakened Galois's genius for mathematics. He raced through the usual texts and went straight for the masters of the day, devouring Adrien Marie Legendre's work on geometry and the memoirs of Joseph Louis Lagrange: *The Resolution of Algebraic Equations, The Theory of Analytic Functions* and *Lessons on the Calculus of Functions*. It was undoubtedly from Lagrange that Galois first learned the theory of equations, to which he was to make fundamental contributions over the next four years. Vernier seems to have appreciated his student's talents: his remarks in Galois's trimester reports carry high praise such as "zeal and success" and "zeal and progress very marked."

With his discovery of mathematics Galois's personality underwent a striking change. He began to neglect his other courses and aroused the hostility of his teachers in the humanities. His rhetoric teachers called him "dissipated" on the trimester reports, and the words "withdrawn," "bizarre" and "original" appear on his evaluations. Even Vernier, although not seeking to cool Galois's passion for mathematics, urged him to work more systematically. Galois did not follow the advice: he decided to take the entrance examination for the École Polytechnique a year early and without the usual preparatory courses in mathematics. Evidently lacking some of the basics, he failed.

Galois considered the failure an injustice, and it hardened his attitude toward authority. Nevertheless, he continued to progress rapidly in mathematics and enrolled in the more advanced course at Louis-le-Grand taught by Louis-Paul-Émile Richard, a distinguished instructor. Richard immediately recognized Galois's abilities and called for his admission to the École Polytechnique without examination. Although the recommendation was not followed, Richard's encouragement produced spectacular results. In March, 1829, while Galois was still a student, his first paper was published. It was titled "Proof of a Theorem on Periodic Continued Fractions" and appeared in *Annales de mathématiques pures et appliquées* of Joseph Diaz Gergonne.

The paper, however, was a minor aside. Galois had already turned to the theory of equations, the topic he had first explored in the works of Lagrange. At age 17 he was taking on one of the most difficult problems in mathematics, a problem that had confounded mathematicians for more than a century.

In 1829 the central question for the theory of equations was: Under what conditions can an equation be solved? More precisely, what was sought was a method of solving an equation having a single variable x whose coefficients are all rational numbers and whose highest-power term is x^n. The method was to be a general one that could be applied to all such equations, and it was to rely on only the four elementary operations of arithmetic (addition, subtraction, multiplication and division) and the extraction of roots. If the solutions or roots of an equation can be obtained from the coefficients solely by these operations, the equation is said to be solvable by radicals.

From a historical perspective it was natural to expect that solving an equation of the nth degree would call for no operations more elaborate than the extraction of nth roots. The solution to the general quadratic, or second-degree, equation $ax^2 + bx + c = 0$, known to the Babylonians, requires the extraction of the square root of a function of the coefficients, namely $b^2 - 4ac$. Hence the general quadratic equation is solvable by radicals. Similarly, the general solution to the cubic equation, devised by the Italian mathematicians Scipione dal Ferro and Niccolò Fontana, or Tartaglia, in the early 1500's, requires taking cube roots of functions of the coefficients. The solution to the general fourth-degree equation, first achieved by the Italian mathematician Lodovico Ferrari at about the same time, requires the extraction of fourth roots.

By the time of Galois, however, nearly 300 years of effort had not yielded a solution by radicals to general equations of the fifth degree or higher. A number of mathematicians had come to suspect that such general solutions are not possible, even though in special cases, such as the equation $x^7 - 2 = 0$, the solution can be found by radicals. (In this instance one solution is $\sqrt[7]{2}$.) Galois provided definitive criteria for determining whether or not the solutions to a given equation can be found by radicals. Per-

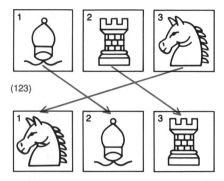

(123)

IDEA OF A GROUP can be illustrated by considering the group $S(3)$, which is the group of permutations of three objects. An element of $S(3)$ operates on the objects by rearranging them. The permutation (123) moves the object in the first square into the second, the object in the second square into the third and the object in the third square into the first. Because there are six possible arrangements of three objects there are six elements in $S(3)$.

haps even more remarkable than Galois's findings in the theory of equations were the methods he developed to study the problem. His investigations led to a theory with applications far outside the theory of equations, now called the theory of groups.

Galois submitted his first papers on what was to become group theory to the French Academy of Sciences on May 25 and June 1, 1829, near the end of his final year at Louis-le-Grand. Less than two months later he was to take the entrance examination to the École Polytechnique for the second time, but meanwhile events in his life took an unfortunate turn. On July 2, a few weeks before the examination, Évariste's father suffocated himself in his Paris apartment. The Jesuit priest of Bourg-la-Reine had forged Mayor Galois's name to a number of malicious epigrams directed at Galois's own relatives. The senior Galois could not bear the scandal. The entrance examination therefore took place under the worst-possible circumstances. Furthermore, Évariste apparently declined to follow the examiner's suggestions for exposition and was failed for a second and final time. The two disasters crystallized his hatred for the conservative hierarchy then ruling France.

MARGINAL NOTE on one of the papers left behind by Galois on the morning of his duel is the most famous of the documents cited in support of the legend that Galois wrote down his ideas on group theory in a single night. The note reads: "There is something to complete in this demonstration. I do not have the time (author's note)." ("Il y a quelque chose à completer dans cette démonstration. Je n'ai pas le temps (Note de l'A.).") According to the familiar account of the life of Galois by Eric Temple Bell, the phrase "I do not have the time" is written frequently in the manuscripts. Actually the page reproduced here is the only place the phrase appears. The rapid handwriting of the note contrasts sharply with the careful and deliberate hand of the body of the text, which suggests that Galois did not write the paper the night before the duel but only edited it. Indeed, the paper had already been submitted to the Academy of Sciences and returned to Galois by Siméon Denis Poisson with suggestions for reworking it.

ainsi $F = \phi V$, et l'on aura
$$\phi V = \phi V' = \phi V'' = \cdots = \phi V^{(n-1)} \quad 4$$
La valeur de V pourra donc se déterminer rationnellement.

2° Réciproquement, et une fonction F est déterminable rationnellement, et que l'on pose $F = \phi V$,
on devra avoir
$$\phi V = \phi V' = \phi V'' = \cdots = \phi V^{(n-1)}$$
puisque l'équation en V n'a pas de diviseur commun avec celle où que V satisfait à l'équation rationnelle $F = \phi V$,
F étant une quantité rationnelle. Donc la fonction F
sera nécessairement invariable par les substitutions
du groupe écrit ci-dessus.

Ainsi ce groupe jouit de la double propriété dont il
s'agit dans le théorème précédent proposé. Le
théorème est donc démontré.

Nous appelons groupe de l'équation le groupe en question.

Scholie. Il est évident que dans le groupe de permutations
dont il s'agit ici, la disposition des lettres n'est point
à considérer, mais seulement les __substitutions__ de lettres
par lesquelles on passe d'une permutation à l'autre.
Ainsi l'on peut se donner arbitrairement une première permutation pourvu que les autres substitutions permutations
s'en déduisent par toujours par les mêmes substitutions
de lettres. Le nouveau groupe ainsi formé jouira évi-
demment des mêmes propriétés que le premier, puisque
dans le théorème précédent, il ne s'agit que des
substitutions que de lettres que l'on peut faire dans les
fonctions.

PROPOSITION II.

Théorème. Si l'on adjoint à une équation donnée la racine r d'une
équation auxiliaire irréductible il
arrivera de deux choses l'une : ou bien le groupe de l'équation ne sera pas changé ; ou bien il se partagera en
p groupes appartenant chacun à l'équation proposée respectivement quand on leur adjoint chacun des racines de
l'équation auxiliaire. 2° ces groupes jouiront de la
propriété remarquable, que l'on passera de l'un à l'autre
en f opérant dans toutes les permutations de première une
même substitution de lettres.

1° si, après l'adjonction de r, l'équation en V dont il
est question plus haut reste irréductible, il est clair que
le groupe de l'équation ne sera pas changé. si au contraire
elle se réduit alors l'équation en V se décomposera en
p facteurs tous de même degré et de la forme
$$f(V, r) \times f(V, r') \times f(V, r'') \times \cdots$$
r, r', r'' $\cdots$ étant les diverses valeurs de r.
Ainsi le groupe de l'équation proposée se décomposera
aussi en p groupes chacun d'eux même nombre de
permutations, puisqu'à chaque valeur de V correspond
une permutation. Ces groupes seront respectivement ceux
de l'équation proposée ; quand on lui adjoindra successi-
vement r, r', r'' $\cdots$

Ce qui caractérise un groupe. On peut partir
d'une des permutations quelconques du groupe.
Scholie. les substitutions sont indépendantes
même du nombre des racines.

Il y a quelque chose à compléter dans cette
démonstration. Je n'ai pas le temps.
(Note de l'A.)

car si l'on divise N entre $f(V, r) = 0$ et $F = 0$
F étant la donnée de proposée, il ne peut arriver
que de deux choses l'une ou le résultat de l'élimination
sera de même degré en V que $f(V, r)$, ou il en
d'un degré moindre ou du moins plus grand.

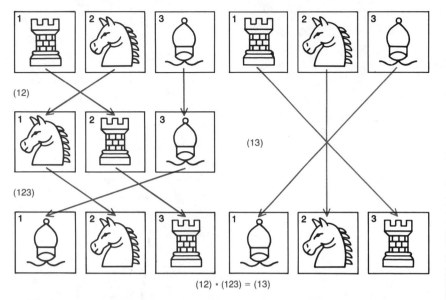

(12)

(13)

(123)

(12) * (123) = (13)

"MULTIPLICATION" of one element in $S(3)$ by another element is carried out by determining the arrangement of objects that results from the operation of the first permutation and then applying the second permutation to this arrangement. The single permutation that brings about the same rearrangement is called the product of the two permutations. In general the multiplication of groups is not commutative: the product of two elements depends on the sequence they are applied. Thus (12)*(123) is equal to (13) but (123)*(12) is equal to (23).

Forced to consider the less prestigious École Normale (then called the École Préparatoire), Galois took the baccalaureate examinations required for entry in November, 1829. This time he passed on the basis of an exceptional score in mathematics, and he was given university status at about the time his first papers on group theory were to be presented to the Academy of Sciences. The papers, however, did not receive a hearing.

When the papers were received by the Academy, Augustin Louis Cauchy, then the most eminent mathematician in France and a staunch supporter of the conservative restoration, was appointed referee. Cauchy had already investigated permutation theory, a forerunner of group theory, and he later wrote extensively on group theory itself. Although legend has it that Cauchy lost, forgot or discarded Galois's manuscripts, it is far more credible that Cauchy recognized their importance and handled them with care. Indeed, a letter discovered in 1971 by Taton in the archives of the Academy makes it clear that on January 18, 1830, Cauchy had planned to give Galois's results a full hearing before the Academy. Cauchy wrote: "I was supposed to present today to the Academy...a report on the work of the young Galois. ... Am indisposed at home. I regret not being able to attend today's session and I should like you to schedule me for the following session for the...indicated subjects."

The following week, however, when Cauchy read a paper of his own to the Academy, he did not present Galois's work. Why this happened remains a subject of speculation. Taton conjectures that Cauchy urged Galois to expand his work and submit it for the Academy's Grand Prize in mathematics. Although Taton's conjecture cannot yet be documented, Galois did submit an entry for the prize in February, the month before the deadline. The entry was sent to Jean Baptiste Joseph Fourier, the mathematician who invented what is now called Fourier analysis, in his capacity as Perpetual Secretary to the Academy. Fourier died in May, however, and Galois's manuscript could not be found among Fourier's effects. Galois later attributed his bad luck to malicious intent on the part of the Academy, accusing the prize committee of rejecting his paper out-of-hand because his name was Galois and because he was still a student. The Galois legend passes on such accusations at face value, but there can be little doubt that Galois's attitude toward authority was becoming somewhat paranoid.

In spite of the setbacks Galois remained a productive mathematician and began to publish in Baron de Férussac's *Bulletin des sciences mathématiques, astronomiques, physiques et chimiques,* a far less conspicuous forum than the meetings of the Academy. His articles make it clear that in 1830 he had progressed beyond all others in the search for the conditions that determine the solvability of equations, although he did not yet have the complete answer in hand. By January, 1831, however, he had reached a conclusion, which he sub-mitted to the Academy in a new memoir, written at the request of the mathematician Siméon Denis Poisson. The paper is the most important of Galois's works, and its existence more than a year before the duel makes nonsense of the story that all Galois's work on the theory of groups was written down in a single night.

In order to understand Galois's work it is unprofitable to study the original papers. Poisson did his best to understand the 1831 manuscript, but he finally recommended that the Academy reject it, encouraging Galois to expand and clarify the exposition. Poisson also criticized one of Galois's proofs as inadequate, although the statement being proved could be shown to be true through a result proved by Lagrange. According to Peter Neumann of the University of Oxford, the criticism is completely accurate. Galois's arguments are presented in a concise form that makes them extremely difficult to follow, and they are not free from error. With the benefit of a century and a half of clarification, however, it is now possible to set forth the essentials of the theory in an accessible form. To this end I have had the assistance of the astrophysicist Adrian C. Ottewill of Oxford.

What is a group? On its deepest level group theory concerns the symmetries inherent in any system. Imagine a snowflake whose points or vertexes are equally spaced at angles of 60 degrees. If the snowflake is rotated about an axis through its center by 60 degrees or any integral multiple of 60 degrees, its basic pattern remains unchanged, even though any particular vertex may change its position. An operation that leaves a pattern invariant in this sense is called a symmetry operation.

If two rotations by integral multiples of 60 degrees are carried out in sequence, the snowflake remains invariant, and the position assumed by the vertexes is one that could have been reached by a single operation. For example, a 60-degree counterclockwise rotation followed by a 240-degree clockwise rotation is equivalent to a 180-degree clockwise rotation. In general, if $R(n)$ denotes a rotation through $60n$ degrees, and if the result of performing first one such operation and then another is written $R(n) * R(m)$, then for all integers n and m the expression $R(n) * R(m)$ is equal to $R(n + m)$. Mathematically the equivalence states that the "product" of two symmetry operations is also a symmetry operation.

There are three other important properties of the rotations of a snowflake. First, a rotation through zero degrees, or $R(0)$, always leaves the pattern invariant, since it does nothing. The product of any rotation $R(n)$ and $R(0)$ is $R(n)$, so that $R(0)$ plays much the same role in

SECOND ELEMENT

	(1)	(123)	(132)	(12)	(13)	(23)
(1)	(1)	(123)	(132)	(12)	(13)	(23)
(123)	(123)	(132)	(1)	(23)	(12)	(13)
(132)	(132)	(1)	(123)	(13)	(23)	(12)
(12)	(12)	(13)	(23)	(1)	(123)	(132)
(13)	(13)	(23)	(12)	(132)	(1)	(123)
(23)	(23)	(12)	(13)	(123)	(132)	(1)

FIRST ELEMENT

MULTIPLICATION TABLE for the six permutations of three objects provides a verification that the permutations satisfy the properties of a group. For every pair of permutations a and b the table shows that their product a * b is itself a permutation. There is an identity element, namely the element (1), with the property that a * (1) is equal to a. For every element a there exists an element called the inverse of a, or a^{-1}, with the property that $a * a^{-1}$ is equal to (1). The inverse of (123), for example, is (132). Finally, the associative law, which states that for any permutations a, b and c the product a * (b * c) is equal to (a * b) * c, can be checked in the table. The permutations in color form a subset of the six permutations. Their multiplication table, also in color, shows that they too form a group. Such a group within a group is called a proper subgroup.

rotations as the number 1 plays in ordinary multiplication. $R(0)$ is therefore called the identity rotation. Second, a rotation $R(n)$ followed by a rotation in the opposite direction by the same amount, which can be denoted $R(-n)$, returns the pattern to its starting point. Thus the product $R(n) * R(-n)$ is equivalent to $R(0)$. The rotation $R(-n)$ is called the inverse of the rotation $R(n)$. Third, the expression $R(m) * R(n) * R(p)$ is unambiguous, because $[R(m) * R(n)] * R(p)$ is equivalent to $R(m) * [R(n) * R(p)]$. This is a formal property of the operation *, by means of which two rotations are combined, called the associative property.

The four properties that hold for combinations of snowflake rotations are characteristic of any set of symmetry operations on any system; they are called the group properties. The system need not be a geometric pattern such as a snowflake. For example, an equation is also a system whose symmetries can be described by the group properties. In abstract terms a group consists of elements or symmetry operations a, b, c and so forth, and a rule denoted by * for combining any two elements. The elements of the group and the rule * are assumed to satisfy the closure criterion, which states that for any elements a and b in the group, a * b is also an element of the group. The group must include an identity element 1, which is defined so that for any element a in the group, a * 1 is equal to a. Moreover, for every element a there must be some inverse element a^{-1} with the property that $a * a^{-1}$ is equal to 1. Finally, the elements of the group and the operation are assumed to satisfy the associative property, which states that $(a * b) * c$ is equal to $a * (b * c)$.

The theory of groups is one of the most fruitful areas of mathematical research; Bell is correct when he writes that it will keep mathematicians busy for hundreds of years. One of the most important recent achievements in group theory was a proof announced at a meeting of the American Mathematical Society in January, 1981, by Daniel Gorenstein of Rutgers University. Gorenstein showed that a list of 26 groups called sporadic finite simple groups is a complete list. In a sense this finding implies that the components or building blocks of any group with a finite number of elements have now been exhaustively classified.

Another set of non-numerical elements that satisfies the properties of a group is the group of permutations on a fixed number of objects. The permuted objects might be chess pieces, for example, or letters of the alphabet. It is essential to recognize, however, that the elements of the group are neither the chess pieces nor the letters but rather the functions that generate the various permutations. To find the "product" of two elements a and b of the group (that is, to find a * b) one finds the result of the first permutation on the set of objects and applies the second permutation to this result.

Suppose three chess pieces are arranged so that a rook is on a square labeled 1, a knight is on a square labeled 2 and a bishop is on a square labeled 3. One element of the permutation group for these objects can be written (12); it takes the object on square 1 and moves it to square 2 and takes the object on square 2 and moves it to square 1. The effect of the element (12) on the arrangement rook-knight-bishop is to exchange the rook and the knight, generating the arrangement knight-rook-bishop. If the operation is then done again, it exchanges the pieces on the same squares, recreating the arrangement rook-knight-bishop. Thus the group element (12) is its own inverse.

Another group element, designated (123), moves the object on square 1 to square 2, the object on square 2 to square 3 and the object on square 3 to square 1. Suppose the initial arrangement rook-knight-bishop is again transformed by the element (12), giving rise to the arrangement knight-rook-bishop. Now the element (123) is applied and generates the arrangement bishop-knight-rook. This final arrangement could have been reached in one step from the initial arrangement by applying the permutation (13), which interchanges the object on square 1 with the object on square 3. Thus the result of the permutation (12) followed by (123) generates the same arrangement of objects as the permutation (13) does. Symbolically, then, $(12) * (123) = (13)$.

The number of permutations or arrangements of n objects is equal to n factorial, written n!. The factorial of a number n is the product of all the whole numbers from 1 to n inclusive; 5!, for

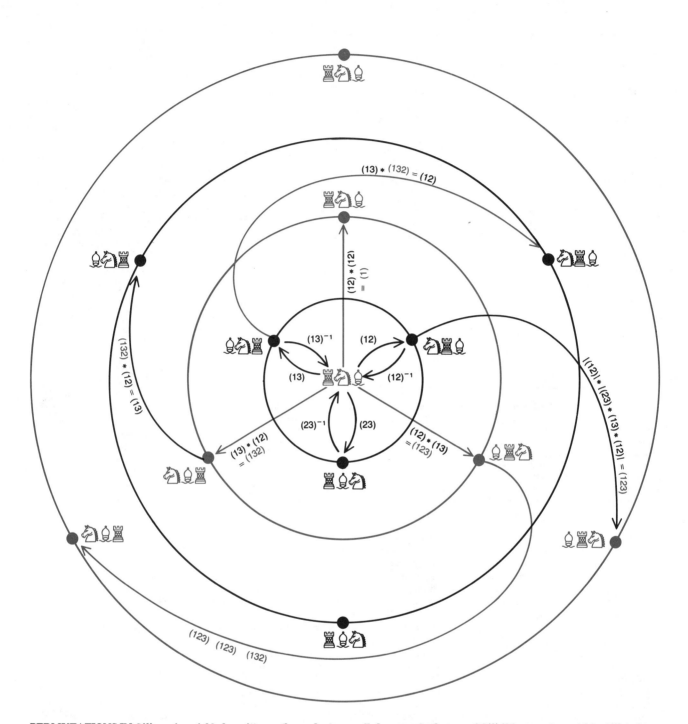

$(13) * {}^{(132)} = (12)$

$(12) * (12)$
$= (1)$

$(132) * (12) = (13)$

$(13)^{-1}$ (12)

(13) $(12)^{-1}$

$(23)^{-1}$ (23)

$(13) * (12)$
$= (132)$

$(12) * (13)$
$= (123)$

$|(12) * |(23) * (13) * (12)|$
$= (123)$

(123) (123) (132)

PERMUTATIONS IN $S(3)$ **can invariably be written as the product of permutations that interchange only two objects. If a permutation can be written as a product of an even number of such interchanges, it is called an even permutation; otherwise it is an odd permutation. If an even permutation (*colored circles*) is multiplied by an even permutation (*colored arrows*), the product is an even permutation; if an even permutation is multiplied by an odd permutation (*black arrows*), the product is odd. Similarly, if an odd permutation (*black circles*) is multiplied by an even permutation, the product is odd, whereas if an odd permutation is multiplied by an odd permutation, the product is even. The even permutations form a subgroup, namely the subgroup in color in the bottom illustration on the opposite page. This subgroup is called the alternating group, or** $A(3)$. **A subgroup such as** $A(3)$ **is called a normal subgroup of** $S(3)$ **if for any element** h **in** $A(3)$ **and any element** g **in** $S(3)$, **the element** $g * h * g^{-1}$ **is also an element of** $A(3)$. **To prove that** $A(3)$ **is a normal subgroup of** $S(3)$ **suppose** g **is an even permutation. Then** $g * h * g^{-1}$ **is the product of three even permutations, which is also an even permutation and so is a member of** $A(3)$. **If** g **is an odd permutation,** $g * h * g^{-1}$ **is the product of an odd by an even by an odd permutation, which is again an even permutation. Hence** $A(3)$ **is a normal subgroup. By a similar argument it can be shown that for any number** n, $A(n)$ **is a normal subgroup of** $S(n)$. **The number of elements in a subgroup must divide the number of elements in the parent group without remainder. Because** $A(n)$ **has half as many elements as** $S(n)$, $A(n)$ **includes the maximum number of elements that a proper subgroup of** $S(n)$ **can have.** $A(n)$ **is the maximal normal subgroup.**

EQUATIONS OF DEGREE 3:
$$ax^3 + bx^2 + cx + d = 0$$

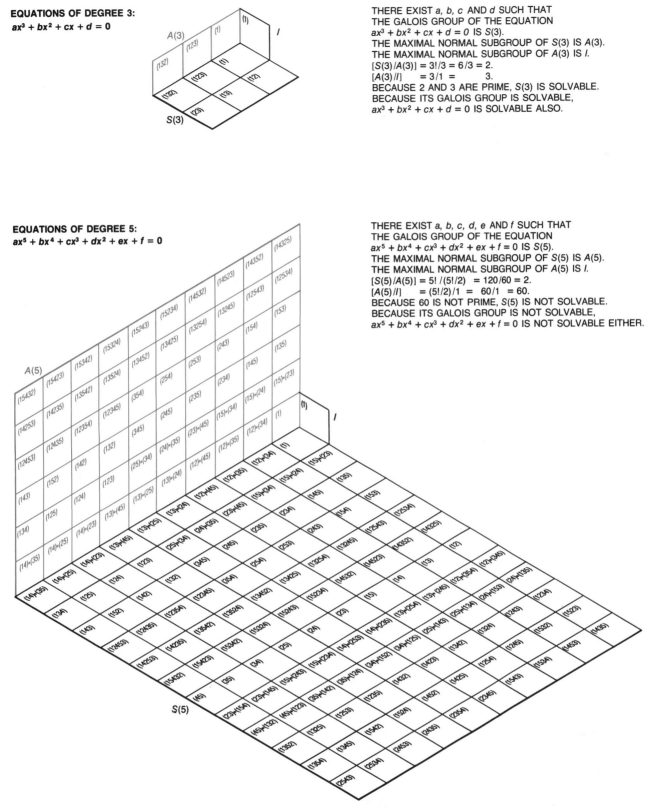

THERE EXIST a, b, c AND d SUCH THAT
THE GALOIS GROUP OF THE EQUATION
$ax^3 + bx^2 + cx + d = 0$ IS $S(3)$.
THE MAXIMAL NORMAL SUBGROUP OF $S(3)$ IS $A(3)$.
THE MAXIMAL NORMAL SUBGROUP OF $A(3)$ IS I.
$[S(3)/A(3)] = 3!/3 = 6/3 = 2$.
$[A(3)/I] = 3/1 = 3$.
BECAUSE 2 AND 3 ARE PRIME, $S(3)$ IS SOLVABLE.
BECAUSE ITS GALOIS GROUP IS SOLVABLE,
$ax^3 + bx^2 + cx + d = 0$ IS SOLVABLE ALSO.

EQUATIONS OF DEGREE 5:
$$ax^5 + bx^4 + cx^3 + dx^2 + ex + f = 0$$

THERE EXIST a, b, c, d, e AND f SUCH THAT
THE GALOIS GROUP OF THE EQUATION
$ax^5 + bx^4 + cx^3 + dx^2 + ex + f = 0$ IS $S(5)$.
THE MAXIMAL NORMAL SUBGROUP OF $S(5)$ IS $A(5)$.
THE MAXIMAL NORMAL SUBGROUP OF $A(5)$ IS I.
$[S(5)/A(5)] = 5!/(5!/2) = 120/60 = 2$.
$[A(5)/I] = (5!/2)/1 = 60/1 = 60$.
BECAUSE 60 IS NOT PRIME, $S(5)$ IS NOT SOLVABLE.
BECAUSE ITS GALOIS GROUP IS NOT SOLVABLE,
$ax^5 + bx^4 + cx^3 + dx^2 + ex + f = 0$ IS NOT SOLVABLE EITHER.

SOLUTION OF EQUATIONS was the problem for which Galois developed group theory. A general method of solution would employ only addition, subtraction, multiplication, division and the extraction of roots, and it could be applied to any equation of degree n, where n is the highest power to which a variable is raised. Galois proved that no such method exists when n is equal to 5 or more. Every equation of degree n can be associated with the group $S(n)$ or with some subgroup of $S(n)$; the group associated with an equation is now called the Galois group of the equation. Galois showed that an equation can be solved by the methods of arithmetic and root extraction only if its Galois group is what he defined as a solvable group. A group is solvable if it generates a series of maximal normal subgroups whose composition factors (determined from the number of elements in the parent group and the subgroups) are prime numbers. The composition factors generated by $S(3)$ and its series of maximal normal subgroups are all prime numbers. It follows that all third-degree equations are solvable. When n is equal to 5 or more, however, it can be proved that the maximal normal subgroup of $A(n)$ is the identity group I, which includes only the identity element. Since $A(n)$ is the maximal normal subgroup of $S(n)$, the composition factors of $S(n)$ when n is equal to 5 or more are not all prime numbers. Hence there are equations of degree 5 or greater that cannot be solved by the allowed methods.

example, is equal to $1 \times 2 \times 3 \times 4 \times 5$, or 120. Hence the number of elements in $S(n)$, the permutation group for n objects, is $n!$. The number of elements in a group is called the "order" of the group. $S(3)$, the permutation group for three objects, includes the 3! (or six) permutations (1), (12), (13), (23), (123) and (132). Here (1) is the identity permutation that leaves any arrangement of objects unchanged.

It turns out that certain subsets of the set of elements in a group can by themselves satisfy all the properties of a group, in which case they are said to form a subgroup. If the number of elements in the subgroup is less than the number of elements in the parent group, the subgroup is called a proper subgroup. For example, it is easy to verify that [(1), (12)] is a group, and so it is a proper subgroup of $S(3)$.

For any proper subgroup H of a group G a number called the composition factor can be defined: it is the order of the parent group divided by the order of the subgroup and is generally written $[G/H]$. The composition factor of the subgroup [(1), (12)] with respect to the group $S(3)$ is 6/2 or 3. According to an elementary theorem of group theory, which I shall not prove here, the order of any subgroup must exactly divide the order of its parent group, so that the composition factor is invariably a whole number.

Galois introduced three critical concepts whose interrelations enabled him to prove that there is no general method for solving an equation of the fifth degree or higher when all solutions must be found by radicals. First Galois noted that every equation can be associated with a group of permutations. Such a group is a representation of the symmetry properties of the equation; it is now called the Galois group.

To appreciate the properties of the Galois group, consider any third-degree equation whose coefficients are rational numbers. It can be proved that such an equation has three roots, although the proof does not reveal whether the roots can be found by radicals. If the roots are designated u, v and w, one can form polynomial functions of them, such as $u - v$ or $uv + w - 1$. Any such function can be converted into a related function by permuting the roots u, v and w. For example, the permutation (12) interchanges u and v and so converts the function $u - v$ into the function $v - u$. Many functions of the roots are changed in value by such a permutation, but some of them are not. For example, the function $u + v + w$ is not changed in value by any permutation of u, v and w. Since the group $S(3)$ includes all possible permutations of u, v and w, $u + v + w$ is said to be invariant under $S(3)$.

It is possible to show that the value of $u + v + w$ is a rational number for any third-degree equation with rational coefficients. Other polynomial functions of the roots may be rational for some equations and irrational for others, depending on the coefficients of the equation. If the value of such a function is rational, there exists some group of permutations of u, v and w that do not change the value of the function. The Galois group of an equation is the largest group of permutations that meet this requirement for every rational-valued polynomial function of the roots. In other words, for any polynomial function of the roots that has a rational value, every permutation in the Galois group leaves the value of the function unchanged. When a permutation of the roots does not change the value of any rational-valued polynomial function of the roots, the roots are indistinguishable for that permutation. Hence the larger the number of elements in the Galois group is, the more permutations there are for which the roots are indistinguishable. For this reason the Galois group is a powerful way of representing the symmetry properties of an equation.

Calculating the Galois group for a given equation is generally difficult, although it can always be done in principle without knowing the values of the roots of the equation. For Galois's purposes, however, the calculation was not necessary. All he needed to show was that there are invariably equations of degree n whose Galois group is the largest-possible group of permutations of the roots, namely $S(n)$.

The second concept introduced by Galois was the concept of a normal subgroup. A subgroup H of a group G is normal in G if and only if the following condition is satisfied: When one "multiplies" any element h of the subgroup H on the left by any element g of the parent group G, and then "multiplies" the product on the right by g^{-1} (the inverse element of g), the result is an element of the subgroup H. Symbolically, if H is normal in G, there is an element h' in H such that $h' = g * h * g^{-1}$. For example, one can verify that [(1), (123), (132)] is a normal subgroup of $S(3)$ [see illustration on page 56].

If a finite group G has any normal subgroups at all, there must be one subgroup whose order is the largest of all the normal subgroups of G; it is called the maximal normal subgroup of G. Similarly, a maximal normal subgroup may in turn have a maximal normal sub-

"INFAMOUS COQUETTE" whom Galois blames for his troubles in a letter written the night before the duel was probably the same woman whose name appears frequently in the margins of Galois's papers. In the manuscript reproduced above the name "Stéphanie" can be read under the name "Évariste," and Galois has also combined the initials "S" and "E" into a single cursive diagram. From letters and other manuscripts it is clear that Galois's angry epithet is his reaction to an unhappy love affair with a woman he had met only a few months before the duel. She has been identified as Stéphanie-Félicie Poterin du Motel, the daughter of a Parisian physician.

group of its own, and the sequence of maximal normal subgroups continues until the smallest normal subgroup possible is reached. Any group G therefore generates a sequence of maximal normal subgroups. If the sequence is labeled $G, H, I, J, \ldots$, then a series of maximal normal composition factors can be defined: $[G/H], [H/I], [I/J]$ and so on.

The third important concept of Galois's theory is the concept of a solvable group. Galois calls a group solvable if every one of the maximal normal composition factors generated by the group is a prime number. The maximal normal subgroup of $S(3)$, for instance, is $[(1), (123), (132)]$. In turn the maximal normal subgroup of $[(1), (123), (132)]$ is $[(1)]$. The composition factor defined for $S(3)$ and its subgroup $[(1), (123), (132)]$ is $6/3$ or 2, and the composition factor for the group $[(1), (123), (132)]$ and its subgroup $[(1)]$ is $3/1$ or 3. Since both 2 and 3 are prime numbers, $S(3)$ is a solvable group.

The term solvable group is well justified by Galois's theory: he was able to show that an equation is solvable by radicals if and only if the Galois group of the equation is a solvable group. In order to prove that equations of the fifth degree or higher cannot in general be solved by radicals, Galois had to show that there are equations of this kind for which the Galois group is not a solvable one. As it happens the group $S(n)$ is not a solvable group when n is equal to 5 or more [*see illustrations on pages 56 and 57*]. Since for all such values of n there are equations of degree n for which $S(n)$ is the Galois group, the general equation of the fifth degree or higher is not solvable.

By the time Galois's work on group theory was nearly completed, the events of his life had become markedly political. In July, 1830, the republican opponents of the restored monarchy took to the streets and the revolution forced the Bourbon king Charles X into exile. While the students of the left-wing École Polytechnique played an active role in the fighting, Galois and his fellows at the École Préparatoire were locked up in the school by its director. Incensed, Galois tried to scale the walls; he failed and thereby missed the brief revolution.

Although the Bourbon abdication seemed a great victory for the republicans, it proved a short-lived one. Louis Philippe was placed on the throne, to the disappointment of Galois and like-minded liberals. In the months following the revolution Galois joined republican societies, met republican leaders (notably François Vincent Raspail) and probably took part in the riots and demonstrations that were racking Paris. He joined the Artillery of the National Guard, a branch of the militia made up

almost entirely of republicans. In December his break with the École Préparatoire became official. He wrote a letter that called the director of the school a traitor for his actions during the July revolution; not surprisingly, Galois was expelled.

The impression of Galois that one gains from the events of this period is not that of a victim of circumstances, as legend would have it. Instead he appears to have been a hothead whose extreme actions consistently got him into trouble. A letter by the mathematician Sophie Germain implies that Galois regularly attended sessions of the Academy of Sciences and habitually insulted the speakers. After his expulsion from the École Préparatoire he moved to his mother's house in Paris but proved to be so difficult to live with that she fled.

The climactic event of the turbulent spring of 1831 took place on May 9 during a republican banquet celebrating the acquittal of 19 artillery officers who had been accused of plotting to overthrow the government. According to the memoirs of Alexandre Dumas (père), Galois stood to propose a toast. "To Louis Philippe!" he said, raising a glass and a dagger at the same time. For this provocative act he was arrested the next day and held for more than a month in the prison at Sainte-Pélagie.

At the ensuing trial Galois's defense claimed that the toast had been "To Louis Philippe, if he betrays" but that "if he betrays" had been drowned out in the uproar. Whether the jurors believed the defense or were moved by Galois's youth (he was then 19) is not known, but they acquitted him in minutes. Nevertheless, on Bastille Day, July 14, 1831, less than a month after his acquittal, Galois was arrested again, this time for illegally wearing the uniform of the Artillery Guard. The guard had been disbanded as a threat to the crown and Galois's gesture was therefore an act of defiance. This time he spent eight months in Sainte-Pélagie.

The prison term was devastating: Galois was alternately despondent and raging. Raspail, who was serving a sentence at the same time, later recalled that Galois once, while drunk, had to be restrained from an attempt at suicide. Later, according to Raspail, Galois confided a chilling vision of his demise: "I shall die in a duel for the sake of some worthless girl [quelque coquette de bas étage]. Why? Because she will invite me to avenge her honor, which another will have compromised." After a fellow prisoner was shot, it seems Galois accused the prison superintendent of arranging the shooting. Galois was subsequently confined to the dungeon, perhaps as a result of the accusation.

In all this turmoil, however, the worst blow was the rejection of Galois's 1831 paper. In the scathing preface to his

memoirs, which he wrote while he was in prison, he stated: "I tell no one that I owe anything of value in my work to his advice or encouragement. I do not say so because it would be a lie."

The end of Galois's life has always had a particular fascination for theorists. Biographers have been unwilling to accept at face value the implication of his own words, namely that the duel was a result of a personal quarrel. Instead his biographers have looked for prostitutes, agents provocateurs and political opponents to account for his death. There is no evidence to support any of these conjectures.

In the middle of March, 1832, Galois was transferred from Sainte-Pélagie to the nursing home Sieur Faultrier because of a cholera epidemic in Paris. There he apparently met the "infamous coquette." The involvement was brief, but it is absurd to suggest that the girl was a prostitute or a conspirator who helped to arrange his assassination. The epithet "infamous coquette" has been associated with the words "quelque coquette de bas étage" and so taken as confirming the prostitute story. According to the account by Raspail, however, the latter phrase was spoken by Galois a year before the duel; it may even have been Raspail's own invention. Moreover, on May 25, six days before his death, Galois alludes to a broken love affair in a letter to his friend, Auguste Chevalier: "How can I console myself when in one month I have exhausted the greatest source of happiness a man can have, when I have exhausted it without happiness, without hope, when I am certain it is drained for life?"

Who was the woman? Two fragmentary letters were written to Galois in the weeks before the duel, suggesting a personal quarrel in which Galois was more of a participant than he admitted. The first letter begins: "Please let us break up this affair. I do not have the wit to follow a correspondence of this nature but I will try to have enough to converse with you as I did before anything happened...." The second letter is similar in tone, and the first of them bear the signature "Stéphanie D." In the Galois manuscripts Carlos Alberto Infantozzi of the University of the Republic in Uruguay has managed to read a name that Galois had erased: Stéphanie Dumotel. Further detective work by Infantozzi shows she was Stéphanie-Félicie Poterin du Motel, the daughter of a resident physician at the Sieur Faultrier. She later married a language professor.

It is also unlikely that the man who killed Galois was in the pay of an antirepublican plot, in spite of the assertion of Galois's brother Alfred that Évariste was murdered. According to Dumas, Galois's adversary was Pescheux d'Herbinville, not a political enemy but an ardent republican. Indeed, d'Herbinville was one of the 19 officers of the Artillery Guard whose acquittal was the occasion of Galois's defiant toast to the king. Moreover, when agents of the crown were exposed during the revolution of 1848, d'Herbinville was not among them. A summary of an article recently sent to me by Taton indicates that the duel was between friends and unfolded as a kind of Russian roulette in which only one pistol was charged.

Galois's mathematical writings the night before the duel were actually confined to making editorial corrections on two manuscripts and to summarizing the contents of these and one other paper in a long letter to Chevalier. The first paper was the one rejected by Poisson; the second was a fragmentary version of an article that had already been published in Férussac's *Bulletin*. The third

has not been found and its content is known only from the summary in the letter; it apparently concerned integrals of general algebraic functions.

What of the famous words "I do not have the time" that Galois is supposed to have written repeatedly in frustration at being unable to complete his work? The phrase does appear, in the margin of the first memoir, but only once. Appended to it in parentheses is the comment "Author's note."

I do not believe that the facts about the life of Évariste Galois as I have presented them diminish his stature as a mathematician in the slightest degree. Many manuscript fragments indicate he carried on his mathematical investigations not only while in prison but also up to the time of his death. That he could work productively through such turbu-

lent times is testimony to the extraordinary fertility of his imagination. Quite apart from the circumstances under which the work was done, there is no question that Galois developed one of the most original ideas in the history of mathematics.

His reputation is not served, however, nor is the history of science, by a legend that insists a scientific genius must be above reproach in his personal life, or that any contemporary who does not appreciate his genius is either a fool, an assassin or a prostitute. The notion that genius is not tolerated by mediocrity is too old a platitude to be adopted uncritically as accurate history. From this point of view a genius would have to be recognized as such even when standing at a banquet table with a dagger in his hand. (*See postscript on page 107.*)

Joseph Henry

by Mitchell Wilson
July 1954

In his lifetime he was famous as a scientific administrator; today he is also known as a great scientist. He discovered induction before Faraday and radio waves long before Hertz

In the spring of 1837 a small group of men in an English laboratory attempted an impromptu experiment: they had rigged up an electric circuit to carry a very feeble current, and they were trying to draw sparks by closing and opening the circuit. Charles Wheatstone touched together the two pieces of wire that completed the circuit. He drew no spark. Michael Faraday said that Wheatstone was going about it in the wrong way. Faraday made a few adjustments and tried his hand. Still no spark.

A visiting American waited patiently while the two famous "electricians" argued back and forth over the probable cause of failure. As the American listened, he absently coiled a length of wire about his finger in a tight corkscrew. After a few minutes, he remarked that, whenever the two gentlemen were ready, he would gladly show them how to draw the spark. Faraday gave him one of his usual brusque answers, but the American went ahead. He added his little coil to one of the leads, and this time, when he opened the circuit, he drew sparks that were clearly visible.

Faraday clapped his hands with delight and said, "Hurrah for the Yankee experiment! What in the world did you do?"

If Joseph Henry had had Faraday's temper, he might have blurted out, "If you would only read what I publish, and understand what you read, you'd know

what you just saw!" Instead the Princeton professor patiently explained· the phenomenon of self-induction to the man whom the world had already credited with the discovery of induction.

There was a century and a quarter of time and a world of knowledge between the electrical experiments of Benjamin Franklin and the electromagnetic theory of James Clerk Maxwell. Much of that knowledge was gathered by one man—Joseph Henry. The time required was only 15 years—1829 to 1844. Yet Henry was a stranger in his own time. His friends mistook his scientific idealism for lack of the American spirit; international science ignored him because he was an American. Not until after he was dead and the contemporaries of his youth were gone did younger men realize that he had been a giant and that the considerable fame he had achieved during the latter half of his life had been for the least of his works. In the end science paid him its greatest tribute by raising him on the pedestal of the lower case: to the electrical units the ampere, the volt, the ohm and the farad was added the henry, the unit of inductance.

During the 25 years before Henry's appearance in science, Alessandro Volta showed how to produce a steady current of electricity, G. S. Ohm found the law that governs the strength of the current and Hans Oersted and Dominique Arago discovered that a current of electricity could create magnetism. Now, in the 1820s, a few clear-headed investigators were pondering the question: If electricity created magnetism, did magnetism in turn create electricity? Joseph Henry, a mathematics teacher in a country school in a provincial town in an undeveloped nation, not only found

the answer, but went far beyond his predecessors in the depth of his research.

In Henry's background there was nothing to indicate either the extent of his ability or the direction his interest would take. He was born in 1797 near Albany and was raised in poverty. Farm hand and storekeeper's apprentice, he was a dreamy boy who barely knew how to read. When he was 13 his main concern was his pet rabbit. One day the rabbit ran away and Henry pursued it by tunneling into a church. He came up inside a locked room which contained a library of romantic novels. He forgot the rabbit and read the books.

He was so enraptured by their melodrama that the next year, when he was sent to Albany to earn a living, the 14-year-old boy made a beeline for the Green Street Theater, where John Bernard was directing his famous company. For two years Joseph Henry was a hard-working, talented apprentice actor.

In his 16th year he made his second great discovery. Confined to his room by illness one day, he happened to pick up a book left by a fellow boarder. Even late in life he could still recall the opening paragraph: "You throw a stone or shoot an arrow into the air, why does it not go forward in a line with the direction you gave it? On the contrary, why does flame or smoke always mount upwards although no force is used to send them in that direction?" Joseph Henry had found the world of science.

Henry was never able to make minor decisions. Once he ordered a pair of shoes and from day to day changed his mind as to whether he wanted square or round toes. The exasperated cobbler made the shoes both ways—one was round-toed, the other square-toed. But Henry made important decisions on the

FORMAL PORTRAIT of Henry was made during the years when he was secretary of the Smithsonian Institution by the famous photographer Mathew Brady. The negative from which this reproduction was made is in the collection of Frederick Hill Meserve.

spur of the moment. With no background, training or tradition he had decided to go on the stage. Now, with even less reason, he abruptly made up his mind to become a natural philosopher.

Henry walked to the Albany Academy and presented himself as a student. The other boys were years younger and the

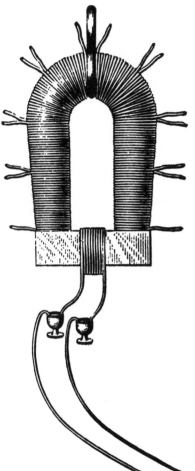

sons of wealthy families, but Henry lived in a private world where everything seemed possible. Fortunately he had so much talent that the real world took the shape of his private dream.

In seven months of night classes and special tutoring he acquired enough learning to get himself an appointment as a country schoolmaster. In this way he could afford to go on with his studies. Teaching and attending classes at the Academy took more than 16 hours a day, but Henry was in love with his life. Later he gave up teaching and talked his professor of chemistry into making him his assistant to set up experiments for public lectures. Henry's theatrical training had taught him that every demonstration must be foolproof, convincing and as dramatic as possible. This experience contributed to the speed and simplicity which later characterized his own experiments.

When Henry had completed his course at the Academy, he took a job as a surveyor and engineer on the Erie Canal. The days of his poverty seemed ended, and the future was wide open. A man of his training could make a fortune almost anywhere from the seaports of the East to the distant hills of Wisconsin. After a few months, however, he was offered the professorship of mathematics and natural philosophy back in Albany. He felt the country needed advanced teachers even more desperately than engineers. Reluctantly he accepted the post.

Joseph Henry rode back to Albany in 1826. At this time he was a young man

of striking appearance: he had curly blond hair, piercing blue eyes and the carriage of an actor. Behind the facade was the basic gift of great investigators—the instinct for reducing ideas to their essential simplicity.

His teaching schedule was heavy; the only time he could steal for research was the summer vacation, when he was permitted to convert one of the classrooms into a laboratory. At the end of August his apparatus was stored away and the benches and desks were returned.

His first work was to build electromagnets along the lines described by William Sturgeon of England. Sturgeon's magnet was a bar of iron coated with shellac, around which was loosely wrapped a length of bare wire. Sturgeon bent his bar into the form of a horseshoe; seven pounds of metal could be lifted into the air when he turned the current on, and just as dramatically dropped when he turned the current off. One summer in the Albany schoolroom Henry built a magnet that could lift one ton. Instead of insulating the iron, Henry had carefully insulated the wire. This allowed him to wrap the wire as closely as he wished, so that he could pack an enormous number of turns along the iron bar. Henry described his device in the *American Journal of Science*, published by Benjamin Silliman of Yale.

The experiments on electromagnets led Henry to the problem of generating electricity from magnetism. All the previous investigators, misled by the fact that a steady electric current induced a steady magnetic field, had sought some arrangement by which a steady magnetic field would induce an electric current. The usual test was to wind a length of wire around a piece of magnetized iron, to rub the free ends of the wire together and to look for sparks. Henry's great achievement was to perceive that the answer lay not in a steady magnetic field, but in a magnetic field that was changing.

In the crucial experiment Henry used one of his horseshoe-shaped electromagnets with a straight piece of soft iron, which he called an armature, running between its poles. Around the armature he wound a length of insulated copper wire about 30 feet long and connected its ends to a galvanometer some 40 feet away. Thus he had two coils completely independent of each other, the magnet coil attached to a battery and the other coil only to the galvanometer. He was ready to begin. "I stationed myself near the galvanometer," he wrote later, "and directed an assistant at a given word to . . . connect the . . . battery attached

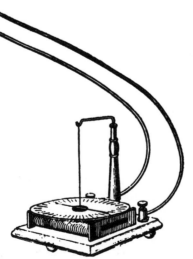

EARLY EXPERIMENT by Henry first made electricity from magnetism. At the top of this illustration is a horseshoe magnet wound with wire. Across its poles is an armature also wound with wire. When Henry turned the magnet current on, a magnetic field was set up in the armature. This induced a current in the armature coil, and the needle of the galvanometer moved. When the current was turned off, the needle swung the other way.

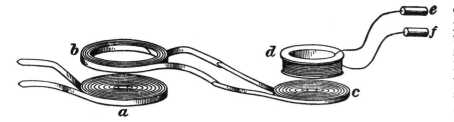

TRANSFORMER principle was described in an early paper by Henry. A coil of flat copper strip (a) was connected to a battery. When the current was interrupted, a current was induced in a second coil (b). If this coil was connected to a third coil (c), a current could also be induced in a fourth coil (d). By holding a pair of electrodes (e and f) in his hands, Henry was able to gauge how much the coils stepped up or stepped down the voltage.

to the magnet." Then the miracle happened. "The north end of the galvanometer needle was deflected 30 degrees, indicating a current of electricity in the wire surrounding the armature."

An instant later Henry must have been disappointed. Even though the current continued to flow through the magnet coil, the needle returned to its zero position. He signaled his assistant to turn the current off. To his amazement the moment the circuit was broken the needle moved again, but in a direction opposite that of the first swing.

Henry soon guessed the reason for this unexpected behavior. It was only while the magnetism in the armature was changing—from zero to its full value as the magnetic circuit was closed, from full value back to zero as the circuit was opened—that anything happened in the secondary coil. He summed up the effect as he understood it in this way: "An instantaneous current in one or the other direction accompanies any change in the magnetic intensity of the iron."

Henry had now established that a current will be induced in any wire in a changing field. He shortly discovered that "any wire" includes the very wire that created the field in the first place. As early as 1829 he had observed the magnetic effect of a current on itself—now called self-induction. It was by making use of this phenomenon that he later confounded Faraday and Wheatstone.

Now this great work, and much more, was done in consecutive summers before 1831; but the first account of it, from which the foregoing quotations are taken, was tragically not written until 1832. Henry knew that he was working on the most difficult problem of his day; he knew that he had solved it before anyone else. But he had never had any personal contact with science as a profession, and the European scientists whose names he knew seemed figures of towering stature. He was therefore reluctant to publish any of his results until he could

accumulate an overwhelming mass of data. His modesty was actually the unconscious pride of genius demanding to be accepted on its own terms. In addition, he was terribly pressed for time. There was not a moment to spare for the laborious work of composition.

For the rest of his life he was to regret that he had not publicized his results. "I ought to have published earlier," he said sadly. "I ought to have published, but I had so little time! It was so hard to get things done! I wanted to get out my results in good form, and how could I know that another on the other side of the Atlantic was busy with the same thing?"

The blow fell in May, 1832. Still filled with the confidence that he was years ahead of the world on a great work, he casually picked up a British journal. He read two paragraphs and the magazine slowly fell from his hands: he was years ahead of nobody. Faraday had just reported his independent discovery of electromagnetic induction.

Faraday's 1832 paper was based on results achieved as recently as the previous autumn. Although Henry had been years ahead of Faraday, he now felt that there was no point in publishing his own results at all. He was sick with despair. However, Silliman had heard of Henry's work and continually pressed him to describe it for the *American Journal of Science*. Henry finally sat down and began the series of papers that was to secure his place in history, but only after his death.

Not since the scientific work of Benjamin Franklin had there been such a chance for American science to achieve world distinction. The young Republic was particularly sensitive to the European attitude that America had nothing culturally to offer. Instead of sympathizing with Henry, many of his friends blamed him for having failed to publish in time. They called him "irresponsible" and "unpatriotic." There were a few who understood. Instead of punishing him, they increased his opportunities for research by getting him an appointment to the faculty at Princeton.

While still at Albany, Henry had invented the electrical relay. He used it to create the first electromagnetic telegraph system, anticipating Samuel F. B. Morse by at least five years. Henry's signaling device was a bell. He never published the details of the relay as a separate paper. He lectured on its practical importance, but to him it was merely an adaptation of the much deeper principles he had already propounded. He explained the device to Morse and to Wheatstone, the inventor of the English telegraph, and both men used it freely.

Henry's relay was a horseshoe magnet

ALBANY ACADEMY was where Henry acquired most of his education. In seven months he was able to learn enough to become a country schoolmaster and continue his studies.

wound with the wire of the long telegraphic sending circuit. Across the pole pieces of the horseshoe was a movable iron armature, which was pulled toward the magnet each time a current impulse of the signal arrived. As the armature moved up and down it mechanically opened and closed a second circuit which contained its own battery. The second circuit contained either a printing mechanism or the horseshoe coil of still another relay so that the strengthened signal could be sent on again. Except for mechanical details Henry's relay has gone unchanged.

At Princeton he built an enlarged telegraphic device and sent signals over a mile of wire, stating that successive relays would allow him to continue the circuit indefinitely. He continued his researches on induction, achieving a remarkable understanding of the details of the phenomenon. In one page he described what was in effect the principle of the electric transformer: "The apparatus used in the experiment consists of a number of flat coils of copper ribbon. . . . Coil No. 1 was arranged to receive the current from a small battery, and coil No. 2 placed on this, with a glass interposed to insure perfect insulation; as often as the circuit of No. 1 was interrupted, a powerful secondary current was induced in No. 2. . . . The shock, however, from this coil is very feeble, and scarcely can be felt above the fingers." In other words, the current had been increased, but the voltage had been stepped down. "Coil No. 1 remaining as before, a longer coil was substituted for No. 2. With this arrangement the magnetizing power was much less, but the shocks were more powerful." Now he had cut down the current, but stepped up the voltage.

Henry's contemporaries knew so little about electricity and electric circuits that they could find in his work only what they were equipped to understand. To those who read the *American Journal of Science*—and its circulation was extremely small—Henry had simply improved the electromagnet. His fundamental insight into the transformer was entirely missed, and therefore forgotten within a few years. Very few Europeans bothered to read the *American Journal*. A decade after the publication of Henry's original papers they were reprinted in England, but even then they were only superficially appreciated.

Henry rarely used mathematics in his analysis of physical phenomena. In his time Ohm's law—today taught to high-school students—had not yet been reduced to quantitative terms. Henry's analysis was powerful, but it was qualitative rather than quantitative. Voltages were given a relative measurement by the intensity of the shock felt by the experimenter; current intensities were similarly measured either by chemical means or, when they were very weak, by the acid taste produced in the experimenter's mouth. Henry detected feeble voltages by the shock to his tongue. But even though he dealt only with relative quantities, he was able to arrive at the correct exponential shape of the growth-and-decay curve for current in an inductive circuit.

Henry made his last great contribution to the study of electricity in 1842. In that year he demonstrated the transmission of radio waves. It was half a century before the celebrated experiments of Heinrich Hertz. Henry noticed that the effect of a spark could be detected by a parallel circuit 30 feet away. Spark coils operating on the second floor of his laboratory building magnetized needles in the basement, the induction taking place through 30 feet of air and two layers of 14-inch flooring. The following excerpt from his paper shows he clearly understood that this was a wave phenomenon, and that it was identical with the propagation of light: "It would appear that the transfer of a single spark is sufficient to disturb perceptibly the electricity of space throughout at least a cube of 400,000 feet of capacity; and when it is considered that . . . the spark is [oscillatory] . . . it may be further inferred that the diffusion of motion in this case is almost comparable with that of a spark from a flint and steel in the case of light."

In 1846 Henry's career as a research worker came to an end. The U. S. Government was seeking a director for the newly founded Smithsonian Institution, and Henry was offered the post. To accept meant that all his time would be devoted to administrative duties. But Henry felt that here was a great opportunity to give American science a cohesive form. Twenty years earlier a sense of duty had caused him to give up a profitable career in engineering. Now he felt it was his duty to give up research in order to act as the first national administrator of science.

By the time Henry was in his fifties, he was considered one of America's leading scientists. But his contemporaries knew him as a scientific administrator: director of the Smithsonian, adviser on science to Abraham Lincoln during the Civil War, the man to whom young inventors like Morse and Alexander Graham Bell went for encouragement. They did not know him as the scientific worker whose 15 years of electromagnetic research was far ahead of its time.

Henry's work as director of the Smithsonian touched many fields. He set up a project to report information about the weather, which later developed into the U. S. Weather Bureau. He persuaded James Lick to found his famous observatory in California. He served on innumerable government advisory boards, including the commission that in the 1850s examined the plans for an ironclad gunboat for the U. S. Navy. Henry was the only commissioner to recommend the design. His advice was disregarded, and, when the Civil War broke out, the design was adopted by the Confederacy in the building of the *Merrimac*.

The meteorological data which Henry collected for the Smithsonian was gathered by telegraph from 500 observers throughout the country east of the Mississippi River. As each telegraphic report came in from a local area, a small round card was pinned in position on a large map of the country. Different colors indicated rain, snow, clear weather or cloudiness. Henry said that storms moved eastward at the rate of 20 to 30 miles per hour, and he successfully taught the usefulness of the weather map to farmers, railroad people and shipping interests.

Henry was the first man to study the relative temperature of sunspots. In 1848 he projected an image of the sun on a white screen. Using a very small thermopile he was able to measure the relative temperature of each point on the projected image. He discovered that the images of the spots were cooler than the areas around them.

The development of the dynamo in the last decade of Henry's life marked the beginning of the use of alternating current. Only then were men able to go back and appreciate the importance of Henry's work. Maxwell's electromagnetic theory of the 1860s pointed up, in retrospect, Henry's statement that the propagation of electricity through space was identical with that of light. Hertz's experiments enabled investigators to look back and understand that Henry himself had been transmitting signals of spark frequency and receiving them on crudely tuned circuits. Henry received full honors only after his death because it took 40 years for men to know enough to appreciate what he had done.

Charles Darwin

by Loren C. Eiseley
February 1956

In 1831 this gentle Englishman set forth on his famous voyage in the Beagle. *After 28 years he published* Origin of Species, *which revolutionized man's view of nature and his place in it*

In the autumn of 1831 the past and the future met and dined in London—in the guise of two young men who little realized where the years ahead would take them. One, Robert Fitzroy, was a sea captain who at 26 had already charted the remote, sea-beaten edges of the world and now proposed another long voyage. A religious man with a strong animosity toward the new-fangled geology, Captain Fitzroy wanted a naturalist who would share his experience of wild lands and refute those who used rocks to promote heretical whisperings. The young man who faced him across the table hesitated. Charles Darwin, four years Fitzroy's junior, was a gentleman idler after hounds who had failed at medicine and whose family, in desperation, hoped he might still succeed as a country parson. His mind shifted uncertainly from fox hunting in Shropshire to the thought of shooting llamas in South America. Did he really want to go? While he fumbled for a decision and the future hung irresolute, Captain Fitzroy took command.

"Fitzroy," wrote Darwin later to his sister Susan, "says the stormy sea is exaggerated; that if I do not choose to remain with them, I can at any time get home to England; and that if I like, I shall be left in some healthy, safe and nice country; that I shall always have assistance; that he has many books, all instruments, guns, at my service. . . . There is indeed a tide in the affairs of men, and I have experienced it. Dearest Susan, Goodbye."

They sailed from Devonport December 27, 1831, in H.M.S. *Beagle*, a 10-gun brig. Their plan was to survey the South American coastline and to carry a string of chronometrical measurements around the world. The voyage almost ended before it began, for they at once encountered a violent storm. "The sea ran very high," young Darwin recorded in his diary, "and the vessel pitched bows under and suffered most dreadfully; such a night I never passed, on every side nothing but misery; such a whistling of the wind and roar of the sea, the hoarse screams of the officers and shouts of the men, made a concert that I shall not soon forget." Captain Fitzroy and his officers held the ship on the sea by the grace of God and the cat-o'-nine-tails. With an almost irrational stubbornness Darwin decided, in spite of his uncomfortable discovery of his susceptibility to seasickness, that "I did right to accept the offer." When the *Beagle* was buffeted back into Plymouth Harbor, Darwin did not resign. His mind was made up. "If it is desirable to see the world," he wrote in his journal, "what a rare and excellent opportunity this is. Perhaps I may have the same opportunity of drilling my mind that I threw away at Cambridge."

So began the journey in which a great mind untouched by an old-fashioned classical education was to feed its hunger upon rocks and broken bits of bone at the world's end, and eventually was to shape from such diverse things as bird beaks and the fused wing-cases of island beetles a theory that would shake the foundations of scientific thought in all the countries of the earth.

The Intellectual Setting

The intellectual climate from which Darwin set forth on his historic voyage was predominantly conservative. Insular England had been horrified by the excesses of the French Revolution and was extremely wary of emerging new ideas which it attributed to "French atheists." Religious dogma still held its powerful influence over natural science. True, the 17th-century notion that the world had been created in 4004 B.C. was beginning to weaken in the face of naturalists' studies of the rocks and their succession of life forms. But the conception of a truly ancient and evolving planet was still unformed. No one could dream that the age of the earth was as vast as we now know it to be. And the notion of a continuity of events—of one animal changing by degrees into another—seemed to fly in the face not only of religious beliefs but also of common sense. Many of the greatest biologists of the time—men like Louis Agassiz and Richard Owen—tended to the belief that the successive forms of life in the geological record were all separate creations, some of which had simply been extinguished by historic accidents.

Yet Darwin did not compose the theory of evolution out of thin air. Like so many great scientific generalizations, the theory with which his name is associated had already had premonitory beginnings. All of the elements which were to enter into the theory were in men's minds and were being widely discussed during Darwin's college years. His own grandfather, Erasmus Darwin, who died seven years before Charles was born, had boldly proposed a theory of the "transmutation" of living forms. Jean Baptiste Lamarck had glimpsed a vision of evolutionary continuity. And Sir Charles Lyell—later to be Darwin's lifelong confidant—had opened the way for the evolutionary point of view by demonstrating that the planet must be very old—old enough to allow extremely slow organic change. Lyell dismissed the notion of catastrophic extinction of animal forms on a world-wide scale as impossible, and he made plain that natural forces—the work of wind and frost and water—were sufficient to explain most of the phenomena found in the rocks, provided these forces were seen as operating over enormous periods. Without Lyell's gift of time in immense quantities, Darwin would not have been able to devise the theory of natural selection.

If all the essential elements of the Darwinian scheme of nature were known prior to Darwin, why is he accorded so important a place in biological history? The answer is simple: Almost every great scientific generalization is a supreme act of creative synthesis. There comes a time when an accumulation of smaller discoveries and observations can be combined in some great and comprehensive view of nature. At this point the need is not so much for increased numbers of facts as for a mind of great insight capable of taking the assembled information and rendering it intelligible. Such a synthesis represents the scientific mind at its highest point of achievement. The stature of the discoverer is not diminished by the fact that he has slid into place the last piece of a tremendous puzzle on which many others have worked. To finish the task he must see correctly over a vast and diverse array of data.

Still it must be recognized that Darwin came at a fortunate time. The fact that another man, Alfred Russel Wallace, conceived the Darwinian theory independently before Darwin published it shows clearly that the principle which came to be called natural selection was in the air—was in a sense demanding to be born. Darwin himself pointed out in his autobiography that "innumerable well-observed facts were stored in the minds of naturalists ready to take their proper places as soon as any theory which would receive them was sufficiently explained."

The Voyage

Darwin, then, set out on his voyage with a mind both inquisitive to see and receptive to what he saw. No detail was too small to be fascinating and provocative. Sailing down the South American coast, he notes the octopus changing its color angrily in the waters of a cove. In the dry arroyos of the pampas he observes great bones and shrewdly seeks to relate them to animals of the present. The local inhabitants insist that the fossil bones grew after death, and also that certain rivers have the power of "changing small bones into large." Everywhere men wonder, but they are deceived through their thirst for easy explanations. Darwin, by contrast, is a working dreamer. He rides, climbs, spends long

days on the Indian-haunted pampas in constant peril of his life. Asking at a house whether robbers are numerous, he receives the cryptic reply: "The thistles are not up yet." The huge thistles, high as a horse's back at their full growth, provide ecological cover for bandits. Darwin notes the fact and rides on. The thistles are overrunning the pampas; the whole aspect of the vegetation is altering under the impact of man. Wild dogs howl in the brakes; the common cat, run wild, has grown large and fierce. All is struggle, mutability, change. Staring into the face of an evil relative of the rattlesnake, he observes a fact "which appears to me very curious and instructive, as showing how every character, even though it may be in some degree independent of structure . . . has a tendency to vary by slow degrees."

He pays great attention to strange animals existing in difficult environ-

ments. A queer little toad with a scarlet belly he whimsically nicknames *diabolicus* because it is "a fit toad to preach in the ear of Eve." He notes it lives among sand dunes under the burning sun, and unlike its brethren, cannot swim. From toads to grasshoppers, from pebbles to mountain ranges, nothing escapes his attention. The wearing away of stone, the downstream travel of rock fragments and boulders, the great crevices and upthrusts of the Andes, an earthquake—all confirm the dynamic character of the earth and its great age.

Captain Fitzroy by now is anxious to voyage on. The sails are set. With the towering Andes on their right flank they run north for the Galápagos Islands, lying directly on the Equator 600 miles off the west coast of South America. A one-time refuge of buccaneers, these islands are essentially chimneys of burned-out volcanoes. Darwin remarks that they

PHOTOGRAPHIC PORTRAIT of Darwin was made some years after the appearance of *Origin of Species.* **It is from the collection of George Eastman House in Rochester, N. Y.**

THREE IMPORTANT FIGURES in the life of Darwin are shown here and on the following page. They appear in *Portraits of Men of Eminence*, three volumes of which were published between 1863 and 1865. This book is also from George Eastman House. At left is Robert Fitzroy, Captain of the *Beagle;* at right, Charles Lyell, the geologist who was Darwin's lifelong confidant.

remind him of huge iron foundries surrounded by piles of waste. "A little world in itself," he marvels, "with inhabitants such as are found nowhere else." Giant armored tortoises clank through the undergrowth like prehistoric monsters, feeding upon the cacti. Birds in this tiny Eden do not fear men: "One day a mocking bird alighted on the edge of a pitcher which I held in my hand. It began very quietly to sip the water, and allowed me to lift it with the vessel from the ground." Big sea lizards three feet long drowse on the beaches, and feed, fantastically, upon the seaweed. Surveying these "imps of darkness, black as the porous rocks over which they crawl," Darwin is led to comment that "there is no other quarter of the world, where this order replaces the herbivorous mammalia in so extraordinary a manner."

Yet only by degrees did Darwin awake to the fact that he had stumbled by chance into one of the most marvelous evolutionary laboratories on the planet. Here in the Galápagos was a wealth of variations from island to island—among the big tortoises, among plants and es-

pecially among the famous finches with remarkably diverse beaks. Dwellers on the islands, notably Vice Governor Lawson, called Darwin's attention to these strange variations, but as he confessed later, with typical Darwinian lack of pretense, "I did not for some time pay sufficient attention to this statement." Whether his visit to the Galápagos was the single event that mainly led Darwin to the central conceptions of his evolutionary mechanism—hereditary change within the organism coupled with external selective factors which might cause plants and animals a few miles apart in the same climate to diverge—is a moot point upon which Darwin himself in later years shed no clear light. Perhaps, like many great men, nagged long after the event for a precise account of the dawn of a great discovery, Darwin no longer clearly remembered the beginning of the intellectual journey which had paralleled so dramatically his passage on the seven seas. Perhaps there had never been a clear beginning at all—only a slowly widening comprehension until what had been seen at first mistily

and through a veil grew magnified and clear.

The Invalid and the Book

The paths to greatness are tricky and diverse. Sometimes a man's weaknesses have as much to do with his rise as his virtues. In Darwin's case it proved to be a unique combination of both. He had gathered his material by a courageous and indefatigable pursuit of knowledge that took him through the long vicissitudes of a voyage around the world. But his great work was written in sickness and seclusion. When Darwin reached home after the voyage of the *Beagle*, he was an ailing man, and he remained so to the end of his life. Today we know that this illness was in some degree psychosomatic, that he was anxiety-ridden, subject to mysterious headaches and nausea. Shortly after his voyage Darwin married his cousin Emma Wedgwood, granddaughter of the founder of the great pottery works, and isolated himself and his family in a little village in Kent. He avoided travel like the plague,

THE THIRD IMPORTANT FIGURE, shown above, is Thomas Huxley, who defended Darwin in debate.

save for brief trips to watering places for his health. His seclusion became his strength and protected him; his very fears and doubts of himself resulted in the organization of that enormous battery of facts which documented the theory of evolution as it had never been documented before.

Let us examine the way in which Darwin developed his great theory. The nature of his observations has already been indicated—the bird beaks, the recognition of variation and so on. But it is an easier thing to perceive that evolution has come about than to identify the mechanism involved in it. For a long time this problem frustrated Darwin. He was not satisfied with vague references to climatic influence or the inheritance of acquired characters. Finally he reached the conclusion that since variation in individual characteristics existed among the members of any species, selection of some individuals and elimination of others must be the key to organic change.

This idea he got from the common recognition of the importance of selective breeding in the improvement of domestic plants and livestock. He still did not understand, however, what selective force could be at work in wild nature. Then in 1838 he chanced to read Thomas Malthus, and the solution came to him.

Malthus had written in 1798 a widely read population study in which he pointed out that the human population tended to increase faster than its food supply, precipitating in consequence a struggle for existence.

Darwin applied this principle to the whole world of organic life and argued that the struggle for existence under changing environmental conditions was what induced alterations in the physical structure of organisms. To put it in other words, fortuitous and random variations occurred in living things. The struggle for life perpetuated advantageous variations by means of heredity. The weak and unfit were eliminated and those with the best heredity for any given environment were "selected" to be the parents of the next generation. Since neither life nor climate nor geology ever ceased changing, evolution was perpetual. No organ and no animal was ever in complete equilibrium with its surroundings.

This, briefly stated, is the crux of the Darwinian argument. Facts which had been known before Darwin but had not been recognized as parts of a single scheme—variation, inheritance of variation, selective breeding of domestic plants and animals, the struggle for existence—all suddenly fell into place as "natural selection," as "Darwinism."

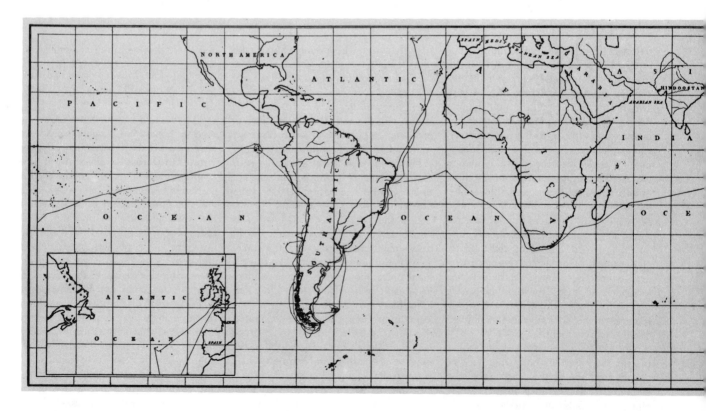

VOYAGE OF THE BEAGLE is traced in this map from Fitzroy's *Narrative of the Surveying Voyage of His Majesty's Ships Adven-* ***ture and Beagle.** The Beagle's course on her departure from and return to England is at lower left. The ship made frequent stops at*

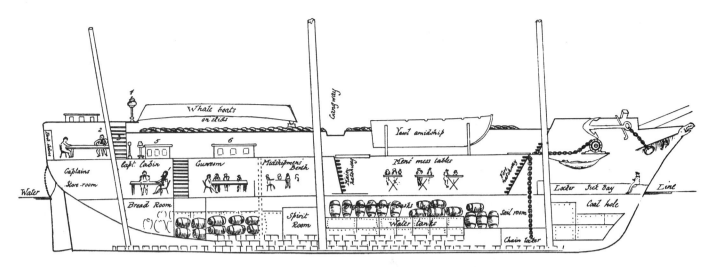

H. M. S. BEAGLE was drawn in cross section many years after the voyage by Philip Gidley King, who accompanied Darwin when he was ashore during the voyage. Darwin is shown in two places: the captain's cabin (*small figure 1 at upper left*) and poop cabin (2).

Procrastination

While he developed his theory and marshaled his data, Darwin remained in seclusion and retreat, hoarding the secret of his discovery. For 22 years after the *Beagle*'s return he published not one word beyond the bare journal of his trip (later titled *A Naturalist's Voyage around the World*) and technical monographs on his observations.

Let us not be misled, however, by

oceanic islands. The Galápagos Islands are on the Equator to the west of South America.

Darwin's seclusiveness and illness. No more lovable or sweet-tempered invalid ever lived. Visitors, however beloved, always aggravated his illness, but instead of the surly misanthropy which afflicts most people under similar circumstances, the result in Darwin's case was merely nights of sleeplessness. Throughout the long night hours his restless mind went on working with deep concentration; more than once, walking alone in the dark hours of winter, he met the foxes trotting home at dawn.

Darwin's gardener is said to have responded once to a visitor who inquired about his master's health: "Poor man, he just stands and stares at a yellow flower for minutes at a time. He would be better off with something to do." Darwin's work was of an intangible nature which eluded people around him. Much of it consisted in just such standing and staring as his gardener reported. It was a kind of magic at which he excelled. On a visit to the Isle of Wight he watched thistle seed wafted about on offshore winds and formulated theories of plant dispersal. Sometimes he engaged in activities which his good wife must surely have struggled to keep from reaching the neighbors. When a friend sent him a half ounce of locust dung from Africa, Darwin triumphantly grew seven plants from the specimen. "There is no error," he assured Lyell, "for I dissected the seeds out of the middle of the pellets." To discover how plant seeds traveled, Darwin would go all the way down a grasshopper's gullet, or worse, without embarrassment. His eldest son Francis spoke amusedly of his father's botanical experiments: "I think he personified each seed as a small demon trying to elude him by getting into the wrong heap, or jumping away all together; and this gave to the work the excitement of a game."

The point of his game Darwin kept largely to himself, waiting until it should be completely finished. He piled up vast stores of data and dreamed of presenting his evolution theory in a definitive, monumental book, so large that it would certainly have fallen dead and unreadable from the press. In the meantime, Robert Chambers, a bookseller and journalist, wrote and brought out anonymously a modified version of Lamarckian evolution, under the title *Vestiges of the Natural History of Creation*. Amateurish in some degree, the book drew savage onslaughts from the critics, including Thomas Huxley, but it caught the public fancy and was widely read. It passed through numerous editions both in England and America—evidence that *sub rosa* there was a good deal more interest on the part of the public in the "development hypothesis," as evolution was then called, than the fulminations of critics would have suggested.

Throughout this period Darwin remained stonily silent. Many explanations of his silence have been ventured by his biographers: that he was busy accumulating materials; that he did not wish to affront Fitzroy; that the attack on the *Vestiges* had intimidated him; that he thought it wise not to write upon so controversial a subject until he had first acquired a reputation as a professional naturalist of the first rank. Primarily, however, the basic reason lay in his personality—a nature reluctant to face the storm that publication would bring about his ears. It was pleasanter to procrastinate, to talk of the secret to a few chosen companions such as Lyell

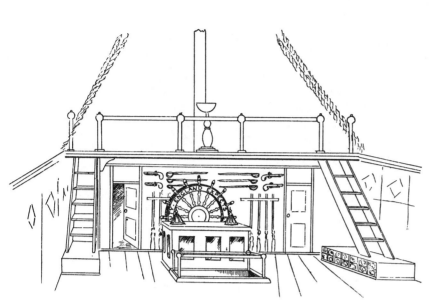

QUARTER-DECK of the *Beagle* is depicted in this drawing by King. In the center is the wheel, the circumference of which is inscribed: "England expects every man to do his duty."

and the great botanist Joseph Hooker.

The Darwin family had been well-to-do since the time of grandfather Erasmus. Charles was independent, in a position to devote all his energies to research and under no academic pressure to publish in haste.

"You will be anticipated," Lyell warned him. "You had better publish." That was in the spring of 1856. Darwin promised, but again delayed. We know that he left instructions for his wife to see to the publication of his notes in the event of his death. It was almost as if present fame or notoriety were more than he could bear. At all events he continued to delay, and this situation might very well have continued to the end of his life, had not Lyell's warning suddenly come true and broken his pleasant dream.

Alfred Russel Wallace, a comparatively unknown, youthful naturalist, had divined Darwin's great secret in a moment of fever-ridden insight while on a collecting trip in Indonesia. He, too, had put together the pieces and gained a clear conception of the scheme of evolution. Ironically enough, it was to Darwin, in all innocence, that he sent his manuscript for criticism in June of 1858. He sensed in Darwin a sympathetic and traveled listener.

Darwin was understandably shaken.

The work which had been so close to his heart, the dream to which he had devoted 20 years, was a private secret no longer. A newcomer threatened his priority. Yet Darwin, wanting to do what was decent and ethical, had been placed in an awkward position by the communication. His first impulse was to withdraw totally in favor of Wallace. "I would far rather burn my whole book," he insisted, "than that he or any other man should think that I had behaved in a paltry spirit." It is fortunate for science that before pursuing this quixotic course Darwin turned to his friends Lyell and Hooker, who knew the many years he had been laboring upon his *magnum opus*. The two distinguished scientists arranged for the delivery of a short summary by Darwin to accompany Wallace's paper before the Linnaean Society. Thus the theory of the two men was announced simultaneously.

Publication

The papers drew little comment at the meeting but set in motion a mild undercurrent of excitement. Darwin, though upset by the death of his son Charles, went to work to explain his views more fully in a book. Ironically he called it *An Abstract of an Essay on the*

Origin of Species and insisted it would be only a kind of preview of a much larger work. Anxiety and devotion to his great hoard of data still possessed him. He did not like to put all his hopes in this volume, which must now be written at top speed. He bolstered himself by references to the "real" book—that Utopian volume in which all that could not be made clear in his abstract would be clarified.

His timidity and his fears were totally groundless. When the *Origin of Species* (the title distilled by his astute publisher from Darwin's cumbersome and half-hearted one) was published in the fall of 1859, the first edition was sold in a single day. The book which Darwin had so apologetically bowed into existence was, of course, soon to be recognized as one of the great books of all time. It would not be long before its author would sigh happily and think no more of that huge, ideal volume which he had imagined would be necessary to convince the public. The public and his brother scientists would find the *Origin* quite heavy going enough. His book to end all books would never be written. It did not need to be. The world of science in the end could only agree with the sharp-minded Huxley, whose immediate reaction upon reading the *Origin* was: "How extremely stupid not to have thought of that!" And so it frequently seems in science, once the great synthesizer has done his work. The ideas were not new, but the synthesis was. Men would never again look upon the world in the same manner as before.

No great philosophic conception ever entered the world more fortunately. Though it is customary to emphasize the religious and scientific storm the book aroused—epitomized by the famous debate at Oxford between Bishop Wilberforce and Thomas Huxley—the truth is that Darwinism found relatively easy acceptance among scientists and most of the public. The way had been prepared by the long labors of Lyell and the wide popularity of Chambers' book, the *Vestiges*. Moreover, Darwin had won the support of the great Hooker and of Huxley, the most formidable scientific debater of all time. Lyell, though more cautious, helped to publicize Darwin and at no time attacked him. Asa Gray, one of America's leading botanists, came to his defense. His codiscoverer, Wallace, as generous-hearted as Darwin, himself advanced the word "Darwinism" for Darwin's theory, and minimized his own part in the elaboration of the theory as "one week to 20 years."

This sturdy band of converts assumed

the defense of Darwin before the public, while Charles remained aloof. Sequestered in his estate at Down, he calmly answered letters and listened, but not too much, to the tumult over the horizon. "It is something unintelligible to me how anyone can argue in public like orators do," he confessed to Hooker, though he was deeply grateful for the verbal sword-play of his cohorts. Hewett Watson, another botanist of note, wrote to him shortly after the publication of the *Origin*: "Your leading idea will assuredly become recognized as an established truth in science, *i.e.*, 'Natural Selection.' It has the characteristics of all great natural truths, clarifying what was obscure, simplifying what was intricate, adding greatly to previous knowledge. You are the greatest revolutionist in natural history of this century, if not of all centuries."

Watson's statement was clairvoyant. Not a line of his appraisal would need to be altered today. Within 10 years the *Origin* and its author were known all over the globe, and evolution had become the guiding motif in all biological studies.

Summing up the achievement of this book, we may say today, first, that Darwin had proved the reality of evolutionary change beyond any reasonable doubt, and secondly, that he had demonstrated, in natural selection, a principle capable of wide, if not universal, application. Natural selection dispelled the confusions that had been introduced into biology by the notion of individual creation of species. The lad who in 1832 had noted with excited interest "that there are three sorts of birds which use their wings for more purposes than flying; the Steamer [duck] as paddles, the Penguin as fins, and the Ostrich (*Rhea*) spreads its plumes like sails" now had his answer—"descent with modification." "If you go any considerable lengths in the admission of modification," warned Darwin, "I can see no possible means of

drawing the line, and saying here you must stop." Rung by rung, was his plain implication, one was forced to descend down the full length of life's mysterious ladder until one stood in the brewing vats where the thing was made. And similarly, rung by rung, from mudfish to reptile and mammal, the process ascended to man.

A Small Place for Man

Darwin had cautiously avoided direct references to man in the *Origin of Species*. But 12 years later, after its triumph was assured, he published a study of human evolution entitled *The Descent of Man*. He had been preceded in this field by Huxley's *Evidences as to Man's Place in Nature* (1863). Huxley's brief work was written with wonderful clarity and directness. By contrast, the *Descent of Man* has some of the labored and inchoate quality of Darwin's overfull folios of data. It is contradictory in spots, as

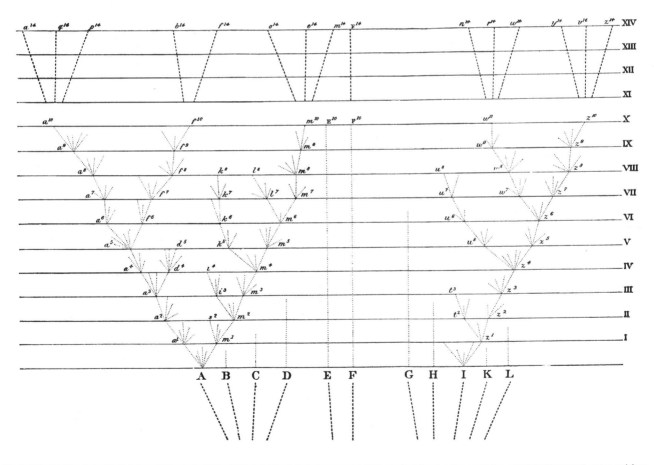

NATURAL SELECTION through the divergence of characters is illustrated in *Origin of Species*. The capital letters at the bottom of the illustration represent different species of the same genus. Each horizontal line, labeled with a Roman numeral at the right, represents 1,000 or more generations. Darwin believed that some of the original species, such as A, would diverge more than others. After many generations they would give rise to new varieties, such as a^1 and m^1. These new varieties would diverge in turn. After thousands of generations the new varieties would give rise to entirely new species, such as a^{14}, q^{14}, p^{14} and so on. The original species would meantime have died out. Darwin thought that only some species of the original genus would diverge sufficiently to give rise to new species. Some of the species, such as F, would remain much the same. Others, such as B, C and D, would die out.

though the author simply poured his notes together and never fully read the completed manuscript to make sure it was an organic whole.

One of its defects is Darwin's failure to distinguish consistently between biological inheritance and cultural influences upon the behavior and evolution of human beings. In this, of course, Darwin was making a mistake common to biologists of the time. Anthropology was then in its infancy. In the biological realm, the *Descent of Man* did make plain in a general way that man was related to the rest of the primate order, though the precise relationship was left ambiguous. After all, we must remember that no one had yet unearthed any clear fossils of early man. A student of evolution had to content himself largely with tracing morphological similarities between living man and the great apes. This left considerable room for speculation as to the precise nature of the human ancestors. It is not surprising that they were occasionally visualized as gorilloid beasts with huge canine teeth, nor that Darwin wavered between this and gentler interpretations.

An honest biographer must record the fact that man was not Darwin's best subject. In the words of a 19th-century critic, his "was a world of insects and pigeons, apes and curious plants, but man as he exists, had no place in it." Allowing for the hyperbole of this religious opponent, it is nonetheless probable that Darwin did derive more sheer delight from writing his book on earthworms than from any amount of contemplation of a creature who could talk back and who was apt stubbornly to hold ill-founded opinions. In any case, no man afflicted with a weak stomach and insomnia has any business investigating his own kind. At least it is best to wait until they have undergone the petrification incident to becoming part of a geological stratum.

Darwin knew this. He had fled London to work in peace. When he dealt with the timid gropings of climbing plants, the intricacies of orchids or the calculated malice of the carnivorous sundew, he was not bedeviled by metaphysicians, by talk of ethics, morals or the nature of religion. Darwin did not wish to leave man an exception to his system, but he was content to consider man simply as a part of that vast, sprawling, endlessly ramifying ferment called "life." The rest of him could be left to the philosophers. "I have often," he once complained to a friend, "been made wroth (even by Lyell) at the confidence with which people speak of the introduction of man, as if they had seen him walk on the stage and as if in a geological sense it was more important than the entry of any other mammifer."

Darwin's fame as the author of the theory of evolution has tended to obscure the fact that he was, without doubt, one of the great field naturalists of all time. His capacity to see deep problems in simple objects is nowhere better illustrated than in his study of movement in plants, published some two years before his death. He subjected twining plants, previously little studied, to a series of ingenious investigations of pioneer importance in experimental botany. Perhaps Darwin's intuitive comparison of plants to animals accounted for much of his success in this field. There is an entertaining story that illustrates how much more perceptive than his contemporaries he was here. To Huxley and another visitor, Darwin was trying to explain the remarkable behavior of *Drosera*, the sundew plant, which catches insects with grasping, sticky hairs. The two visitors listened to Darwin as one might listen politely to a friend who is slightly "touched." But as they watched the plant, their tolerant poise suddenly vanished and Huxley cried out in amazement: "Look, it *is* moving!"

The Islands

As one surveys the long and tangled course that led to Darwin's great discovery, one cannot but be struck by the part played in it by oceanic islands. It is a part little considered by the general public today. The word "evolution" is commonly supposed to stand for something that occurred in the past, something involving fossil apes and dinosaurs, something pecked out of the rocks of eroding mountains—a history of the world largely demonstrated and proved by the bone hunter. Yet, paradoxically, in Darwin's time it was this very history that most cogently challenged the evolutionary point of view. Paleontology was not nearly so extensively developed as today, and the record was notable mainly for its gaps. "Where are the links?" the critics used to rail at Darwin. "Where are the links between man and ape—between your lost land animal and the whale? Show us the fossils; prove your case." Darwin could only repeat: "This is the most obvious and gravest objection which can be urged against my theory. The explanation lies, as I believe, in the extreme imperfection of the geological record." The evidence for the continuity of life must be found else-

where. And it was the oceanic islands that finally supplied the clue.

Until Darwin turned his attention to them, it appears to have been generally assumed that island plants and animals were simply marooned evidences of a past connection with the nearest continent. Darwin, however, noted that whole classes of continental life were absent from the islands; that certain plants which were herbaceous (nonwoody) on the mainland had developed into trees on the islands; that island animals often differed from their counterparts on the mainland.

Above all, the fantastically varied finches of the Galápagos particularly amazed and puzzled him. The finches diverged mainly in their beaks. There were parrot-beaks, curved beaks for probing flowers, straight beaks, small beaks—beaks for every conceivable purpose. These beak variations existed nowhere but on the islands; they must have evolved there. Darwin had early observed: "One might really fancy that, from an original paucity of birds in this archipelago, one species had been taken and modified for different ends." The birds had become transformed, through the struggle for existence on their little islets, into a series of types suited to particular environmental niches where, properly adapted, they could obtain food and survive. As the ornithologist David Lack has remarked: "Darwin's finches form a little world of their own, but one which intimately reflects the world as a whole" [see the article "Darwin's Finches," by David Lack, SCIENTIFIC AMERICAN, February 1956].

Darwin's recognition of the significance of this miniature world, where the forces operating to create new beings could be plainly seen, was indispensable to his discovery of the origin of species. The island worlds reduced the confusion of continental life to more simple proportions; one could separate the factors involved with greater success. Over and over Darwin emphasized the importance of islands in his thinking. Nothing would aid natural history more, he contended to Lyell, "than careful collecting and investigating of *all the productions* of the most isolated islands. . . . Every sea shell and insect and plant is of value from such spots."

Darwin was born in precisely the right age even in terms of the great scientific voyages. A little earlier the story the islands had to tell could not have been read; a little later much of it began to be erased. Today all over the globe the populations of these little worlds are vanishing, many without ever having

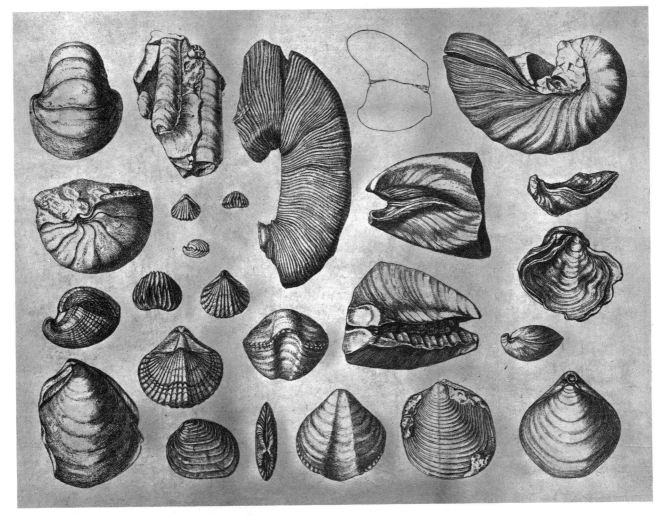

FOSSIL SHELLS were depicted in this engraving from Darwin's *Geological Observations on the Volcanic Islands and Parts of South* *America Visited during the Voyage of H. M. S. Beagle.* This was a technical work published by Darwin before *Origin of Species.*

been seriously investigated. Man, breaking into their isolation, has brought with him cats, rats, pigs, goats, weeds and insects from the continents. In the face of these hardier, tougher, more aggressive competitors, the island faunas—the rare, the antique, the strange, the beautiful— are vanishing without a trace. The giant Galápagos tortoises are almost extinct, as is the land lizard with which Darwin played. Some of the odd little finches and rare plants have gone or will go. On the island of Madagascar our own remote relatives, the lemurs, which have radiated into many curious forms, are now being exterminated through the destruction of the forests. Even that continental island Australia is suffering from the decimation wrought by man. The Robinson Crusoe worlds where small castaways could create existences idyllically remote from the ravening slaughter of man and his associates are about to pass away forever. Every such spot is now a potential air base where the cries of birds are drowned in the roar of jets, and the crevices once frequented by bird life are flattened into the long runways of the bombers. All this would not have surprised Darwin, one would guess.

Of Darwin's final thoughts in the last hours of his life in 1882, when he struggled with a weakening heart, no record remains. One cannot but wonder whether this man who had no faith in Paradise may not have seen rising on his dying sight the pounding surf and black slag heaps of the Galápagos, those islands called by Fitzroy "a fit shore for Pandemonium." None would ever see them again as Darwin had seen them—smoldering sullenly under the equatorial sun and crawling with uncertain black reptiles lost from some earlier creation. Once he had cried out suddenly in anguish: "What a book a devil's chaplain might write on the clumsy, wasteful, blundering, low and horribly cruel works of nature!" He never spoke or wrote in quite that way again. It was more characteristic of his mind to dwell on such memories as that Eden-like bird drinking softly from the pitcher held in his hand. When the end came, he remarked with simple dignity, "I am not in the least afraid of death."

It was in that spirit he had ventured upon a great voyage in his youth. It would suffice him for one more journey.

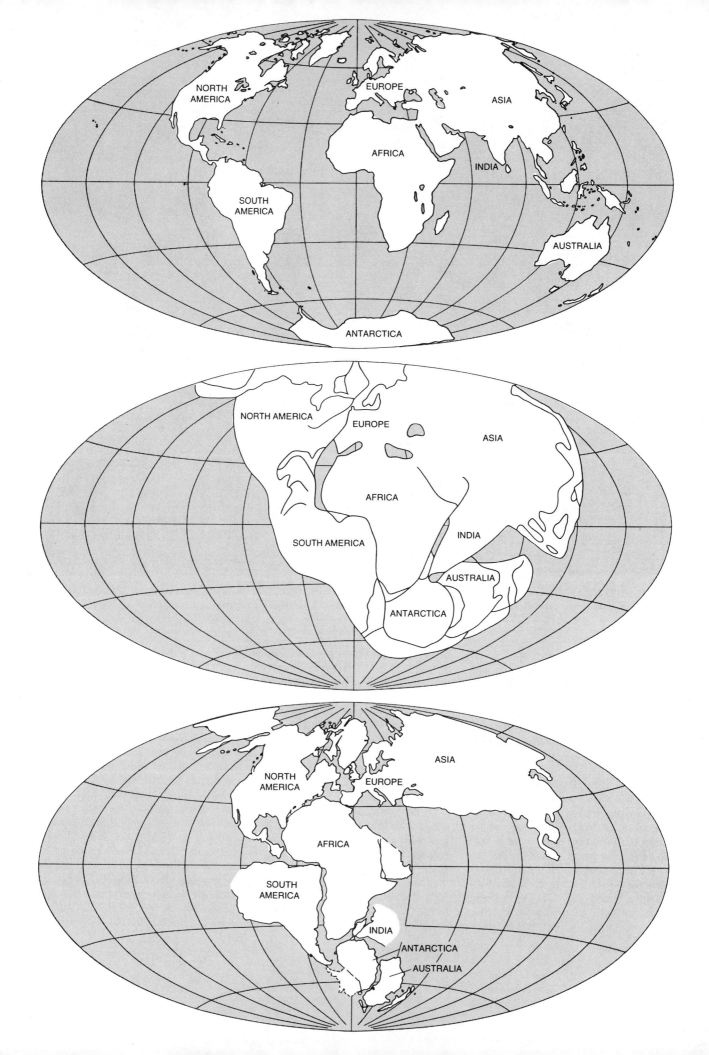

10

Alfred Wegener and the Hypothesis of Continental Drift

by A. Hallam
February 1975

Sixty years ago a German scientist argued that the continents move, and he proposed a history of their migrations. The validity of his theory was not recognized until new evidence emerged in the 1960's

The jigsaw-puzzle fit of the coastlines on each side of the Atlantic Ocean must have been noticed almost as soon as the first reliable maps of the New World were prepared. The complementary shapes of the continents soon provoked speculation about their origin and history, and a number of early theories suggested that the shapes were not the product of mere coincidence. In 1620 Francis Bacon called attention to the similarities of the continental outlines, although he did not go on to suggest that they might once have formed a unified land mass. In the succeeding centuries several other proposals attempted to account for the correspondence, usually as a result of some postulated catastrophe, such as the sinking of the mythical Atlantis. The first to suggest that the continents had actually moved across the surface of the earth was Antonio Snider-Pellegrini in 1858, but he too attributed the event to a supernatural agency: the Great Flood. Today, of course, the migration of the continents is an essential feature of the theory of the earth's structure that is all but universally accepted by geologists.

Of the various hypotheses that preceded the modern theory of plate tectonics, one version stands out: the one propounded by Alfred Wegener in the early years of the 20th century. Wegener had access to only a small part of the in- formation available today, yet his theory anticipates much that is now fundamental to our conception of the earth, including not only the movement of the continents but also the wandering of the poles, with consequent changes in climate, and the significance of the distribution of ancient plants and animals. There are also parts of his work that have turned out to be wrong, but the most important points in his arguments have been substantiated. His hypothesis stands not merely as a forerunner of the concept that now prevails but as its true ancestor.

Wegener first presented his ideas to the scientific community in 1912, but it was not until 50 years later that they gained general currency. When his view of the earth did replace the older model (in the 1960's), the change represented a radical revision of a well-established doctrine, and it took place only because new evidence, derived from discoveries in geophysics and oceanography, compelled it. In the interim Wegener's theory had at best been neglected, and it had often been scorned. At the nadir proponents of continental drift were dismissed contemptuously as cranks. To understand that reaction we must examine both Wegener's work and the attitudes of his contemporaries. Did Wegener derive his conclusions from re- liable evidence, and did he support them with coherent argument? Or did he merely guess, and happen to guess correctly? If his reasoning was plausible, why was his work opposed with such determination and persistence? I shall attempt to answer questions such as these, and in addition to gauge what kind of man Wegener was and to estimate his rank as a scientist.

Wegener had no credentials as a geologist. Born in Berlin in 1880, the son of an evangelical minister, he studied at the universities of Heidelberg, Innsbruck and Berlin and took his doctorate in astronomy. (His most useful accomplishment in the field of his degree was a paper on the Alphonsine tables of planetary motion.) From his early days as a student he had cherished an ambition to explore in Greenland, and he had also become fascinated by the comparatively new science of meteorology. In preparation for expeditions to the Arctic he undertook a program of long walks and learned to skate and ski; in the pursuit of his other avocation he mastered the use of kites and balloons for making weather observations. He was so successful as a balloonist that in 1906, with his brother Kurt, he established a world record with an uninterrupted flight of 52 hours.

Wegener's preparations were rewarded when he was selected as meteorologist to a Danish expedition to northeastern Greenland. On his return to Germany he accepted a junior teaching position in meteorology at the University of Marburg and within a few years had written an important text on the thermodynamics of the atmosphere. A second expedition to Greenland, with J. P. Koch

ANCIENT SUPERCONTINENT incorporated all the earth's large land masses. Wegener's reconstruction of the supercontinent, which he called Pangaea, is shown in the middle illustration on the opposite page. A more recent version, shown at the bottom, differs in details of placement and orientation but preserves the major features of Wegener's proposal. Both maps are based not on the coastlines of the continents but on the edges of the continental shelves. For comparison a map of the world as it appears today is shown at top.

ALFRED WEGENER was by profession a meteorologist; he also participated in three exploratory expeditions to Greenland. In geology and geophysics he was virtually without credentials; nevertheless, it was through his work in these fields that he made his most significant contribution. This drawing of him is based on a photograph made in the 1920's.

his predecessors had: by noting on a map the complementarity of the Atlantic coastlines. By Wegener's own account the notion first occurred to him in 1910, but a contemporary who knew Wegener as a student maintains that he had shown interest in the matter as early as 1903. Whether the idea had been maturing for a decade or for only two years, it was first presented publicly in January, 1912, in a lecture before the German Geological Association in Frankfurt am Main. The first published reports appeared later that year in two German journals.

The prevailing theory of the structure and evolution of the earth in 1912 could not accommodate drifting continents. Geologists and geophysicists then believed that the earth had been formed in a molten state and that it was still solidifying and contracting. During the process the heavy elements, such as iron, had sunk to the core and lighter ones, such as silicon and aluminum, had risen to the surface to form a rigid crust.

To most geologists of the time the model seemed quite successful in accounting for the more prominent features of the earth's surface. Mountain ranges were produced by compression of the surface during contraction, much as wrinkles develop in the skin of a drying, shrinking apple. On a larger scale the pressure generated by contraction, applied through great arches, caused some regions of the surface to collapse and subside, creating the ocean basins, while other areas remained emergent as continents. Vertical movements of the crust were considered entirely plausible, although movements parallel to the surface were excluded. Thus the continents and the ocean basins were in the long run interchangeable; some continental areas sank faster than the adjacent land and were inundated by the sea; at the same time parts of the ocean floor emerged to form dry land.

The similarity or identity of numerous fossil plants and animals on distant continents was explained by postulating land bridges that had once connected the land masses but had since sunk to the level of the ocean floor. The stratification of sedimentary deposits suggested successive marine transgressions onto the continents and regressions from them. The regressions could be attributed to the subsidence of the ocean basins and the transgressions to the partial filling of the basins with sediment eroded from the continents. At about the time Wegener was devising his hypothesis of continental drift a refinement of the traditional view was proposed in which the

of Denmark, followed in 1912. It included the longest crossing of the ice cap ever undertaken; the published glaciological and meteorological findings fill many volumes.

In 1913 Wegener married Else Köppen, the daughter of the meteorologist Wladimir Peter Köppen. After World War I (in which Wegener served as a junior officer) he succeeded his father-in-law as director of the Meteorological Research Department of the Marine Observatory at Hamburg. In 1924 he accepted a chair of meteorology and geophysics at the University of Graz in Austria, where he found that his colleagues were more sympathetic to his research interests than his colleagues in Hamburg had been.

Wegener died while leading a third expedition to Greenland in 1930, probably as a result of a heart attack induced by overexertion. His laudatory obituaries suggest that he had achieved considerable distinction both as a meteorologist and as an Arctic explorer; other sources suggest that in addition he had been a capable organizer and administrator and a lucid and stimulating teacher. His work on continental drift, which will surely be his permanent legacy, had remained a peripheral interest, albeit one that had absorbed him deeply.

Just how Wegener first conceived the idea that the continents could move is not certain. One unauthenticated account has it that he was inspired during a trip to Greenland while watching the calving of glacier ice (the process by which icebergs are born). From his own writings and those of his contemporaries, however, it seems more likely that he came to the theory in the same way that

vertical movements of the crust are governed by isostasy: the concept that all elements of the system are in hydrodynamic equilibrium. Thus the continents, being less dense than the layer under them, float above the ocean floor.

Wegener detected a number of flaws and contradictions in the contracting-earth model. Moreover, many distinctive features of the earth's surface could not be explained at all by that model, unless they were to be considered the result of coincidence. The most obvious of these features is the correspondence between the Atlantic coasts of Africa and South America. (In plotting the correspondence Wegener employed not the coastline itself but the edge of the continental shelf, which is a more meaningful boundary. The same practice is followed in modern reconstructions.) Another anomaly is the distribution of mountain ranges, which are mainly confined to narrow, curvilinear belts; if they had been produced by the contraction of the globe, they should have been spread uniformly over the surface, as the wrinkles on a dried apple are.

Still another peculiarity was discovered in a statistical analysis of the earth's topography. From calculations of the total area of the earth's surface at each of many land elevations and ocean depths, Wegener found that a large fraction of the earth's crust is at two distinct levels. One corresponds to the surface of the continents, the other to the abyssal sea floor [see illustrations on pages 14 and 15]. Such a distribution would be expected in a crust made up of two layers, the upper one consisting of lighter rock, such as granite, and the substratum consisting of basalt, gabbro or peridotite, which would also form the ocean floor. This interpretation is supported by measurements of local variations in the earth's gravity. It is not consistent with a model of the crust in which variations in elevation are the result of random uplift and subsidence; in that case one would expect a Gaussian, or bell-shaped, distribution of elevations around a single median level.

Wegener also found support for his arguments in fossils and distinctive geological features that seemed to cross continental boundaries. In the fossil record an excellent example is provided by the reptiles. Fossils of *Mesosaurus*, a small reptile that lived late in the Paleozoic era, about 270 million years ago, are found in Brazil and in South Africa and nowhere else in the world [see the article "Continental Drift and the Fossil Rec-

ord," by A. Hallam, beginning on page 186]. The peculiar distribution was traditionally explained by the sinking of a land bridge that was assumed to have connected the continents. On geophysical grounds Wegener rejected this explanation; it violated the principle of isostasy, since the material of the bridge would be less dense than that of the sea floor and could not sink into it. The only reasonable alternative was that the continents had once been joined and had since drifted apart.

The geological evidence is of a similar

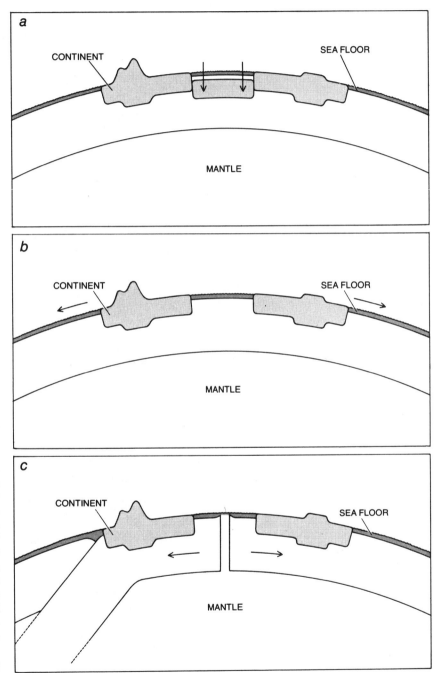

THEORY OF THE EARTH'S STRUCTURE held by most geologists at the beginning of the 20th century was challenged by Wegener. In the traditional view (a) the continents were fixed in place and lateral movement was impossible. Vertical motion of the crust, as in the sinking of land bridges, was allowed in some versions of the theory but objected to in others as a violation of the principle of isostasy, which holds that the continents float in hydrodynamic equilibrium on a substratum of denser material. In Wegener's hypothesis (b) the continents migrate through the substratum, which acts as a viscous fluid. He suggested that they were propelled by forces related to the earth's rotation, but that idea was soon discredited. In the modern theory (c) the continents are carried as passengers on large rigid plates. The plates are forced apart where material wells up to form new sea floor.

nature. For example, large blocks of particularly ancient rock are found both in Africa and across the Atlantic in South America; if the continents are brought together in the proper orientation, the blocks line up precisely [*see illustration on opposite page*]. Wegener himself recognized and described the power of this discovery: "It is just as if we were to refit the torn pieces of a newspaper by matching their edges, then check whether or not the lines of print run smoothly across. If they do, there is nothing left but to conclude that the pieces were in fact joined this way. If only one line was available for the test, we would still have found a high probability for the accuracy of fit, but if we have n lines, this probability is raised to the nth power."

One further line of evidence on which Wegener relied should be mentioned. Geodetic observations made early in the 20th century seemed to indicate that Greenland was moving westward, separating from Europe at a measurable rate. Such a movement might constitute a direct validation of continental drift, but it has not been confirmed in recent measurements employing more accurate techniques.

In order to resolve these contradictions Wegener formulated a comprehensive theory of the origin of the conti-

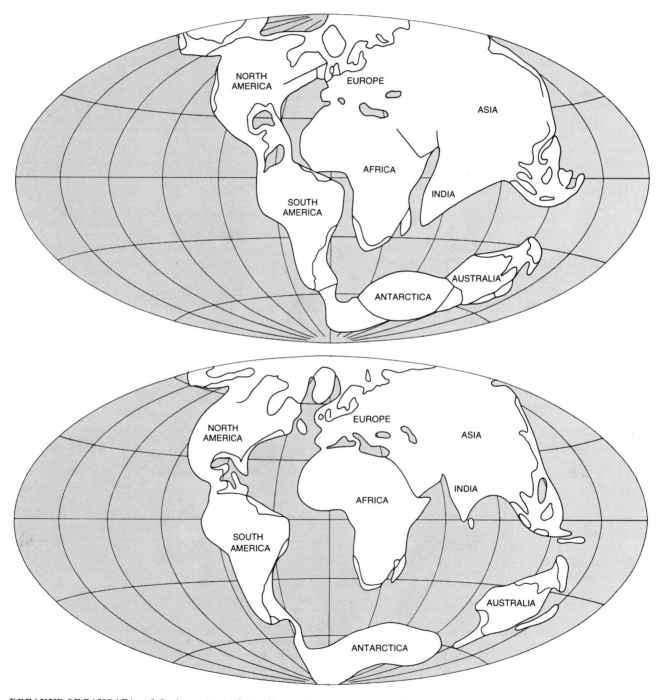

BREAKUP OF PANGAEA and the formation of the modern continents were described by Wegener in two stages. In his account all the continents remained contiguous as late as the Eocene epoch, about 50 million years ago (*upper map*). Even in the early Quaternary period, about two million years ago, South America and Antarctica were connected by an isthmus (*lower map*). Modern workers have considerably revised Wegener's sequence and his dating of events. The breakup actually began about 200 million years ago.

nents. In his reconstruction of earth history all the world's land area was originally united in a single primordial supercontinent, which he named Pangaea from the Greek meaning "all land" [*see illustration on page 76*]. He published his conclusions and evidence in 1915 in a book titled *Die Entstehung der Kontinente und Ozeane* (*The Origin of Continents and Oceans*).

The geophysical basis of Wegener's theory was closely related to the principle of isostasy. Both assume that the substratum underlying the continents acts as a highly viscous fluid; Wegener assumed further that if a land mass could move vertically through this fluid, it should also be able to move horizontally, provided only that a sufficiently powerful motive force was supplied. As evidence that such forces exist he cited the horizontal compression of folded strata in mountain ranges. There is also elegant evidence of the fluid nature of the underlying material: the earth is an oblate sphere, bulging slightly at the Equator, and the size of the bulge is exactly what would be expected for a sphere of a perfect fluid spinning at the same rate. It is a fluid of a special nature. Under short-term stresses, such as those of an earthquake, it acts as an elastic solid; only over the much longer periods of geologic time can its fluid characteristics be observed. Its behavior is analogous to that of pitch, a material that shatters under a hammer blow but flows plastically under its own weight, that is, under the milder but persistently imposed force of gravity.

After World War I Wegener merged his two principal research interests by investigating, with Köppen, changes in world climate through geologic time. By mapping the distribution of certain kinds of sedimentary rock he was able to infer the position of the poles and the Equator in ancient times. His most impressive results were obtained for the Carboniferous and Permian periods, about 300 million years ago [*see illustration on page 84*]. The position of the South Pole was determined from the disposition of boulder beds called tillites, which are formed during the movements of glaciers. In Wegener's reconstruction of Pangaea the pole was just east of what is today South Africa and within ancient Antarctica.

Ninety degrees from the pole Wegener found abundant evidence of a humid equatorial zone. The evidence consists of the vast deposits of coal that stretch from the eastern U.S. to China; fossil plants identifiable within the coals are of a tropical type. Other climatic indi-

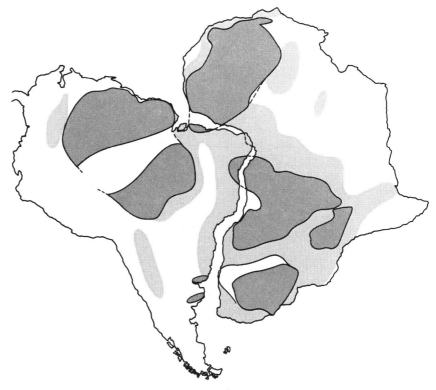

GEOLOGICAL EVIDENCE that the continents once formed a single land mass is provided by distinctive rock formations that can be assembled into continuous belts when Africa and South America are lined up next to each other. Areas shown in dark gray are ancient blocks called cratons; light gray areas represent regions of somewhat younger rock. Wegener considered continuities such as these good evidence for the continental-drift hypothesis.

cators in sedimentary rock are salt and wind-deposited sand, which suggest the presence of ancient deserts. In the past, as today, deserts were formed mainly in the trade-wind belts on each side of the Equator. In the present Northern Hemisphere the Carboniferous coals are replaced in more recent strata by salt and sand-dune deposits that Wegener interpreted as signifying a southeastward shift in the position of the Equator. The movement was confirmed by a corresponding southeastward shift in the main center of tillite deposits, implying that the pole also shifted in that direction.

The last edition of *The Origin of Continents and Oceans* devoted a chapter to ancient climate and contained an extensive discussion of the wandering of the poles. Polar movement is a phenomenon quite distinct from continental drift, but unless the continents are reassembled more or less as Wegener proposed, the distribution of tillites and coal and salt deposits cannot be interpreted coherently.

The last edition also contains more extensive documentation than the earlier ones of similarities in the geology of the southern continents, the outcome of a productive exchange of ideas between

Wegener and the South African geologist Alexander L. du Toit. The essential geophysical arguments, however, remain remarkably similar to those proposed in Wegener's first paper, written almost two decades earlier.

The initial reaction of the scientific community to Wegener's hypothesis was not uniformly hostile, but at best it was mixed. At the first lecture in Frankfurt am Main some geologists were provoked to indignation; at Marburg a few days later, however, the audience seems to have been more sympathetic. Following the early publications several prominent German geologists announced their opposition to "continental displacement" (a more accurate rendering of Wegener's term, *Verschiebung*, than "drift"). A number of geophysicists, on the other hand, expressed approval of the concept. Indeed, in 1921 Wegener was able to say that he knew of no geophysicist who opposed the drift hypothesis.

The early publications, including the first edition of *The Origin of Continents and Oceans*, do not seem to have been read much outside Germany; it was not until the third edition was published in 1922 and translated into several other

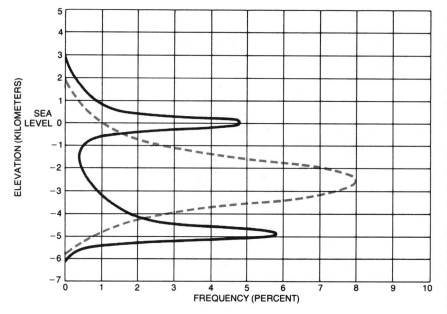

STATISTICAL ANALYSIS of the elevations of the earth's surface provided Wegener with an argument in support of his hypothesis. If topographical features were formed by random uplift and subsidence, a random distribution centered on the mean elevation (*colored line*) would be expected. Actually two levels predominate (*gray line*). One, near sea level, represents average elevation of the continental platforms, the other that of the abyssal sea floor.

languages (including English) two years later that Wegener's hypothesis attracted an international audience. As in Germany, his work was initially given a fair hearing, or at the least it was not dismissed out of hand. At a meeting of the British Association for the Advancement of Science in 1922 discussion of continental drift was "lively but inconclusive," according to the published report. As one would expect, many were skeptical, but the general reception accorded the hypothesis was sympathetic. At about that time several leading geologists on each side of the Atlantic declared themselves in favor of the theory.

The stubborn antagonism to Wegener's ideas that was to be the orthodoxy of geophysics until the 1960's began to develop in the mid-1920's. Two events were instrumental in this hardening of resistance. One was the publication of a treatise titled *The Earth* by Harold Jeffreys of the University of Cambridge; the other was a symposium held in 1928 by the American Association of Petroleum Geologists.

Jeffreys attacked Wegener's theory at what was perhaps its weakest point: the nature of the forces to which Wegener attributed the movement of the continents. Wegener proposed that the westward drift of the Americas could be explained as a consequence of tides in the earth's crust; the postulated force responsible for the northward migration of India and the compression of the crust in the Alpine and Himalayan mountain belts he called *Polflucht,* or "flight from the poles." Jeffreys was able to demonstrate by simple calculations that the earth is far too strong to be even slightly deformed by such forces. If it were not, mountain ranges would collapse under their own weight and the sea floor would be perfectly flat. If the tidal force were strong enough to shift the continents westward, it would also be strong enough to halt the earth's rotation within a year.

These objections are cogent ones, and the mechanism Wegener proposed has long since been abandoned. (The modern theory attributes the movement to the spreading of the sea floor along a system of mid-ocean ridges, where molten rock from the earth's interior wells up.) We can now see, however, that Jeffreys had not refuted the theory merely by demonstrating the inadequacies of the hypothetical motive force. For the most part he simply ignored the empirical evidence, which was the most substantial part of Wegener's argument. Jeffreys dismissed the wandering of the poles as being geophysically impossible.

Of the participants in the American Association of Petroleum Geologists symposium most were hostile, in varying degrees, to Wegener's theory; only one was strongly sympathetic. The proceedings of the symposium are in the main a chorus of criticism. The supposed jigsaw-puzzle fit of the Atlantic continents was inaccurate, the contributors contended, and it did not allow for vertical movements of the crust. The similar rock formations on opposite sides of oceans were not that closely related after all; in any case present similarity did not necessarily imply former contiguity. Ancient animals could have migrated across land bridges. The Carboniferous and Permian tillites of South Africa and other areas were probably not glacial and the Northern Hemisphere coals were probably not tropical. The evidence for the movement of Greenland was inconclusive.

Some of the contributors also demanded that the hypothesis resolve paradoxes that have been approached only in the more recent versions of the theory. For example, they asked why, if the American continents could move laterally by displacing the ocean floor, they crumpled on their western edge, forming the Cordilleran mountain ranges. Did not the compressive force that formed these mountains suggest considerable resistance from the supposedly fluid sea floor? Moreover, why did Pangaea remain intact for most of the earth's history, then abruptly split apart in a few tens of millions of years?

Finally, in addition to questioning Wegener's interpretations and conclusions, some participants in the symposium assailed his credentials and his methods. He was a mere advocate, they protested, selecting for presentation only those facts that would favor his hypothesis. He "took liberties with our globe" and "played a game in which there are no restrictive rules and no sharply drawn code of conduct."

Wegener did not campaign to defend his theory from these criticisms. A middle-aged man with only limited time for research, he felt that he was unable to keep up with the swelling volume of literature and was content to leave the field to younger workers. He did, however, gently rebuke his critics for their partiality. "We are like a judge confronted by a defendant who declines to answer," he said of scientists in relation to the earth, "and we must determine the truth from circumstantial evidence. All the proof we can muster has the deceptive character of this type of evidence. How would we assess a judge who based his decision on only part of the available data?"

After Wegener's death geologists and geophysicists became even more hostile to his account of earth history. In the U.S. the reaction was particularly strong; for an American geologist to express sympathy for the idea of continental drift was for him to risk his career.

Ironically, just as the condemnatory verdict was reached, the theory was significantly strengthened by the contributions of du Toit and of Arthur Holmes of the University of Edinburgh. They eliminated some of Wegener's weaker arguments, marshaled more evidence in support of the hypothesis and provided a more plausible motive mechanism. Holmes suggested that the continents are moved by convection currents in the earth's mantle. His hypothesis was not entirely satisfactory, but it successfully circumvented the criticisms of Jeffreys and his followers. Moreover, it anticipated the modern explanation of why continents move.

In hindsight, and with the knowledge that many aspects of Wegener's theory have been confirmed in the past two decades, we can readily appreciate his accomplishments. He examined critically a model of the earth that was then almost universally accepted. When he discovered weaknesses and inconsistencies, he was bold enough and independent enough to embrace a radical alternative. Furthermore, he had sufficient breadth of knowledge to seek out and perceptively evaluate supporting evidence from a variety of disciplines. The same qualities of mind were applied to the explication of ancient climate and the wandering of the poles.

The intellectual rigor that Wegener brought to his work is illustrated in his own writings and attested to in the words of those who knew him well. In a letter to Köppen written in 1911 Wegener defended his views on continental drift. The letter is reproduced in the biography of Wegener written by his widow. The following passage, which I have translated, has not to my knowledge been published previously in English.

"You consider my primordial continent to be a figment of my imagination, but it is only a question of the interpretation of observations. I came to the idea on the grounds of the matching coastlines, but the proof must come from geological observations. These compel us to infer, for example, a land connection between South America and Africa. This can be explained in two ways: the sinking of a connecting continent or separation. Previously, because of the unproved concept of permanence, people have considered only the former and have ignored the latter possibility. But the modern teaching of isostasy and more generally our current geophysical ideas oppose the sinking of a continent because it is lighter than the material on which it rests. Thus we are forced to consider the alternative interpretation. And if we now find many surprising simplifications and can begin at last to make real sense of an entire mass of geological data, why should we delay in throwing

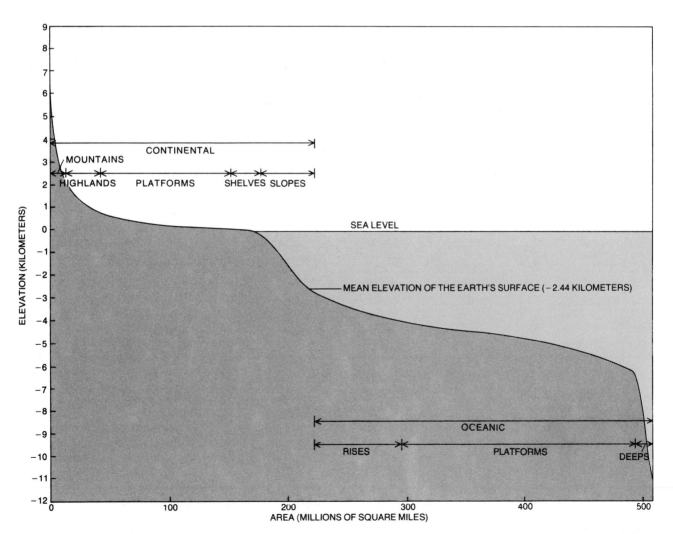

DISTRIBUTION OF ELEVATIONS, when correlated with the area of the earth's surface, confirms that the crust has two fundamental levels. Approximately 300 million square miles of the 510 million square miles of the earth's surface is continental platform or oceanic platform. The distribution conforms to Wegener's model of the earth, in which the continents float in a dense substratum.

the old concept overboard? Is this revolutionary? I don't believe that the old ideas have more than a decade to live. At present the notion of isostasy is not yet thoroughly worked out; when it is, the contradictions involved in the old ideas will be fully exposed." It evidently seemed quite obvious to Wegener, but it is clear that he underestimated the ar-

dor of those committed to the "old ideas."

Wegener's success in constructing from scattered and seemingly unrelated observations a systematic theory of earth history could be attributed to his broad attack on the problem, and perhaps even to his status as a nonspecialist. A glimpse of his approach to scientific problems is provided by an obituary

written by Hans Benndorf, a professor of physics and a colleague of Wegener's at Graz. This passage too was translated by me and has not appeared before in English.

"Wegener acquired his knowledge mainly by intuitive means, never or only quite rarely by deduction from a formula, and when that was the case, it needed

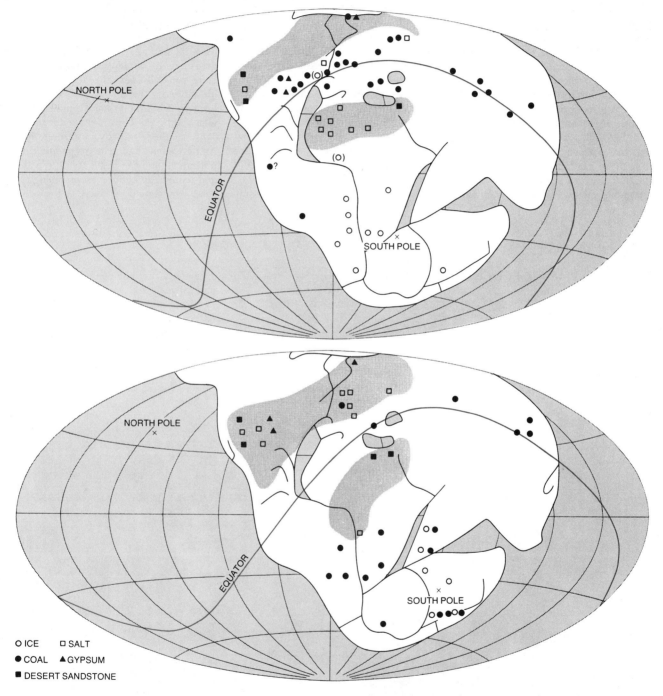

○ ICE □ SALT
● COAL ▲ GYPSUM
■ DESERT SANDSTONE

WANDERING OF THE POLES was proposed by Wegener to account for the distribution of ancient climates. Evidence of tropical climate was provided by certain kinds of coal, of polar climate by the tillites that signal glaciation, and of arid climate by deposits of salt, gypsum and desert sandstone. The symbols employed in the maps are identified in the key at left; in addition arid zones, which are characteristic of trade-wind latitudes, are indicated by gray areas. The upper map shows Wegener's reconstruction for the Carboniferous period, about 300 million years ago; the lower map is for the Permian period, about 230 million years ago. (The distortion of the Equator is caused by the projection employed.) The movement of the continents and the movement of the poles are unrelated, but a coherent account of ancient climate can be devised only by rearranging the continents approximately as Wegener did.

only to be quite simple. Also, if matters concerning physics were involved, that is, in a field distant from his own field of expertise, I was often astonished by the soundness of his judgment. With what ease he found his way through the most complicated work of the theoreticians, with what feeling for the important point! He would often, after a long pause for reflection, say, 'I believe such and such,' and most times he was right, as we would establish several days later after rigorous analysis. Wegener possessed a sense for the significant that seldom erred."

Benndorf's assessment of Wegener's method is supported by the remarks of Wilhelm Max Wundt, who knew Wegener as a student in Berlin: "Alfred Wegener started out to tackle his scientific problems with only quite ordinary gifts in mathematics, physics and the other natural sciences. He was never, throughout his life, in any way reluctant to admit that fact. He had, however, the ability to apply those gifts with great purpose and conscious aim. He had an extraordinary talent for observation and for knowing what is at the same time simple and important, and what can be expected to give a result. Added to this was a rigorous logic, which enabled him to assemble rightly everything relevant to his ideas."

If Wegener was in fact the talented and perceptive scientist his contemporaries describe, and if his conclusions were well rooted in evidence and argument, an obvious question arises: Why was the opposition to his ideas so strong, widespread and persistent?

One possible explanation is that Wegener's theory was "premature" at the time he presented it. Gunther S. Stent of the University of California at Berkeley has argued that an idea must be considered premature if it cannot be connected by a series of simple, logical steps to the canonical, or generally accepted, knowledge of the time [see "Prematurity and Uniqueness in Scientific Discovery," by Gunther S. Stent; this SCIENTIFIC AMERICAN Reader, pg. 95.] A related principle, formulated by Michael Polanyi, a British writer on the philosophy and sociology of science, holds that in science there must always be a prevailing opinion of the nature of things, against which the truth of all assertions is tested. Any observation that seems to contradict the established view of the world must be presumed to be invalid and set aside in the hope that it will eventually turn out to be false or irrelevant. This interpretation of how science works suggests that geologists and geophysicists had to be overwhelmed by evidence, as they were in the 1960's, before they could abandon the established doctrine of stationary continents. Wegener's innovations, precisely because they were innovative, had to be held in abeyance until a new orthodoxy could be created from them.

There is almost certainly truth in this analysis, and reassurance that the rejection of Wegener's work was a necessary circumstance in the orderly progress of science; nevertheless, the analysis is not entirely convincing. It could account for indifference to the hypothesis of continental drift but not for the attitude of the many scientists who relegated the theory to the realm of fantasy. It also fails to explain why the traditional model of the earth was retained even after Wegener demonstrated that there were contradictions in it and even though those contradictions were never resolved. Paleontologists continued to rely on vanishing land bridges, for example, at the same time that geophysicists, who had adopted the principle of isostasy, insisted that the sinking of such bridges was impossible.

It has been suggested that the principal impediment to the acceptance of continental drift was the lack of a plausible motive force after Jeffreys had refuted Wegener's initial proposals. If that is the case, however, why was Holmes's convection-current theory given so little consideration? Furthermore, even today the nature of the motor that moves the continents remains uncertain, yet plate tectonics is so well established that those who reject its basic tenets are commonly dismissed as reactionaries.

Perhaps the long travail of Wegener's hypothesis can be best explained as a consequence of inertia. A geologist at the 1928 symposium of the American Association of Petroleum Geologists is reported to have said: "If we are to believe Wegener's hypothesis, we must forget everything that has been learned in the past 70 years and start all over again." It should also be remembered that to the geologists of the time Wegener was an outsider; they must have regarded him as an amateur. Today, of course, we can see that his position was an advantage because he had no stake in preserving the conventional viewpoint. Moreover, we can see that he was not an amateur after all but an interdisciplinary investigator of talent and vision who surely qualifies for a niche in the pantheon of great scientists.

11

Robert A. Millikan

by Daniel J. Kevles
January 1979

*A tireless investigator and a Nobel prizewinner at a time
when not many Americans were, he had a penchant for
controversy in subjects ranging from cosmic rays (which
he named) to the support of science*

Robert A. Millikan was the most famous American scientist of his day. In 1923 he became the second American (A. A. Michelson had been the first, in 1907) to win the Nobel prize in physics. Millikan is best known to physicists for measuring the charge of the electron with his oil-drop experiment; in the span of a remarkably productive career he also made significant contributions to the study of the photoelectric effect, hot-spark spectra and, above all, cosmic rays. He was more than a research scientist; between the wars he headed the new California Institute of Technology, advised industrial corporations and philanthropic foundations and played a key part in the development of Federal policy for academic science.

The recognition of Millikan's wide-ranging importance has dimmed with the passage of time, in spite of the publication in 1950 of his 300-page autobiography. As such books go, Millikan's is reasonably informative, yet like many autobiographies written at an advanced age (Millikan was 82 when his book was published) it attends more to the earlier events of his life than to the later ones. More important, it omits controversial episodes that were of considerable significance in his career. The controversies centered on cosmic rays and on his attitude toward the relation between academic science and the Government. In both controversies Millikan was on the losing side.

No autobiographer (for that matter no human being) likes to dwell on failure. Millikan went further: he disliked mere suggestions, let alone demonstrations, that he was wrong. Even after he had won the Nobel prize he bridled when the Berkeley physicist Raymond T. Birge proposed that the accuracy of his value for the electron charge might be improved. Edwin C. Kemble of Harvard remarked to Birge: "It's too bad that Millikan isn't a little more of a good sport. A man in his position could well

afford to be." Whether Millikan was sporting or not, his false starts and wrong turns, his commitments to the losing side and his defense of ultimately indefensible redoubts reveal no less than his triumphs the significance of his life both as a research scientist and as a public figure.

The son of a Congregational minister and the former dean of women of a small college, Millikan was born in 1868 in Illinois and was raised from the 1870's in Maquoketa, Iowa (population 3,000). He showed no particular scientific inclinations, and neither his family or school nor the agrarian environment stimulated him to move in a technical direction. At Oberlin College (where his mother had gone before him) he pur-

sued a standard classical curriculum; his move toward physics came when his professor of Greek, impressed with Millikan's abilities, invited him to teach an introductory physics course in the preparatory school run by the college. (When Millikan protested that he knew nothing about the subject, the professor replied, "Anyone who can do well in my Greek can teach physics.") After graduation Millikan stayed at Oberlin to teach in the preparatory department for two more years and then went on to Columbia University, where he was the only graduate student in physics. One summer he worked under Michelson at the University of Chicago. Having earned his doctorate in physics at Columbia, Millikan spent a postdoctoral year in Europe, where his teachers included

PORTRAITS OF MILLIKAN show him (*left to right*) **as an undergraduate at Oberlin College in
the 1880's; in 1891, when he was graduated from Oberlin and stayed on to teach physics in the**

Max Planck, Walther Nernst and Henri Poincaré; he acquired what was on the whole a better than average training for an aspiring American physicist at the turn of the century.

Joining the University of Chicago faculty in 1896, Millikan poured his considerable energies into developing the physics curriculum. At that time American physics students in both high school and college relied on foreign textbooks. Millikan wrote or coauthored a variety of books and laboratory manuals that became classroom standards. (*A First Course in Physics,* written with Henry G. Gale, sold more than 1.6 million copies between 1906 and 1952.) Recognizing, however, that at the University of Chicago the main rewards were given for research rather than for teaching, in about 1908 he put aside the writing of textbooks to concentrate on his work in the laboratory.

World War I interrupted Millikan's research career. By now one of the nation's leading physicists, he was increasingly active in the professional affairs of his discipline and in the National Academy of Sciences. He was also a pioneer in developing links between industry and academic physics: he became a consultant to the research department of Western Electric, primarily to advise the company on vacuum-tube problems, and he pointed a number of his students toward careers in industry. Early in 1917, to help mobilize science for defense, Millikan went to Washington as a vice-chairman and the director of research of the newly established National Research Council in the National

Academy of Sciences. Like many scientists engaged in defense research during World War I, Millikan soon entered the armed services. As a lieutenant colonel in the Army Signal Corps he directed work in meteorology, aeronautical instruments and communications, and in his National Research Council capacity he played an important role in initiating and advancing a major project to develop devices for the detection of submarines.

Millikan's success in the wartime mobilization was no mean achievement. The National Research Council, a private organization like its parent the National Academy, had no Federal appropriation; it had limited private resources and no authority in governmental affairs. Thus disadvantaged in dealing with the Federal bureaucracy, Millikan found that his Army rank did little to ease the difficulty. He was a reserve officer and his leverage with Regular Army officers and career civil servants left a good deal to be desired; to achieve his aims he had to rely heavily on tact, influence and persistence. During the war, as a friend recalled, Millikan learned how to "sell science" to a wide variety of people, military and civilian alike.

Millikan's work in the wartime National Research Council particularly impressed the astrophysicist George Ellery Hale and the physical chemist Arthur A. Noyes, who were inaugurating a venture in education and research in southern California. In 1921, at their urging, Millikan moved to Pasadena to become head (as chief of the executive council) of the new and munificently financed Cal Tech and director of its physics laboratory. ("Just imagine," Wilhelm Röntgen exclaimed, "Millikan is said to have a hundred thousand dollars *a year* for his researches!") Millikan brought scientific acumen and entrepreneurial enthusiasm to the job. He recruited the best faculty possible (including, in spite of a streak of anti-Semitism that was not untypical among academics of the day, such Jewish scientists as Paul S. Epstein and J. Robert Oppenheimer). With a knack for making wealthy southern Californians think it a privilege to be invited to contribute to the institute, he enlarged the Cal Tech endowment and physical plant. Millikan's traits, fused with Hale's vision, Noyes's wisdom and all that money, made Cal Tech virtually an overnight success.

Between the wars Millikan was an influential member of the National Academy and of the National Research Council and its fellowship board, which he helped to administer so as to improve the quality of American physics, particularly in theoretical studies. During World War II he turned over an increasing fraction of his administrative responsibilities at Cal Tech to younger staff members who were running various defense projects. Relinquishing his chief executive's position to Lee A. Du-Bridge in 1946, Millikan remained at the institute until his death in 1953. Throughout the Cal Tech years, in spite of his administrative commitments, he taught a course in atomic physics and took a keen interest in graduate students. He also maintained an active research program almost to the end.

college's preparatory school; as a member of the physics faculty at the University of Chicago soon after the turn of the century; in 1918, when he had made a name as an investigator, and toward the end of his long career as an investigator and an administrator, in about 1940.

The investigation for which Millikan is best known, the oil-drop experiment, was undertaken in Chicago in about 1909, soon after he put aside textbook writing to concentrate on research. His research on the electron charge actually began not with oil but with water. The original idea, based on the work of the British-born physicist H. A. Wilson, was first to measure the rate at which a charged small cloud of water vapor fell under the influence of gravity and then to measure its modified rate of fall under the counterforce of an electric field. Since the mass of the cloud could in principle be determined from Stokes's law of fall, one could calculate the total charge—again in principle. As Millikan recognized, the technique was fraught with uncertainties, including the fact that evaporation at the surface of the cloud confused the measurement of its rate of fall. To correct for this effect he decided to study the evaporation history of the cloud while it was held stationary by a strong electric field.

When Millikan switched on the field, the cloud disappeared, leaving in its place a few charged water droplets moving in a slow, stately manner in response to the imposed electrical force. Millikan quickly realized that he could determine the electron charge accurately by observing the movement of individual charged droplets within an electric field. With this technique Millikan arrived at a rather good value for the electron charge, e, which he soon improved by substituting for water a less volatile oil. Since the accuracy of his value was no better than the constants involved in the calculation, he painstakingly reevaluated the coefficient of viscosity of air and the mean-free-path term in the Stokes-Cunningham version of the law of fall. In 1913 he published his value for the electron charge: $4.774 \pm .009 \times 10^{-10}$ electrostatic unit. It was to serve the world of physics for more than a generation, until it was redetermined more accurately in 1928 by Erik Baecklin, who arrived, by an indirect method involving X rays and crystals, at a value of $4.8033 \pm .002 \times 10^{-10}$ electrostatic unit.

No less important than the number itself was the clear-cut implication that could be drawn from the experiment concerning the nature of the electron. Before Millikan the ratio of the electron's charge to its mass was well established as a constant, but the constancy of the ratio did not demonstrate that all electrons were identical particles; some antiatomistic European physicists had been maintaining that the electron was not a unique particle and that its observed charge was actually the statistical average of diverse electrical energies. Even in the water-drop stage of his experiment, however, Millikan observed that his measured charge was always an

IN UNIFORM AS A LIEUTENANT COLONEL in the Army Signal Corps, Millikan saluted in front of a monument in Washington with his youngest son, Max, in 1918. Millikan was also vice-chairman and director of research of the newly established National Research Council.

integral multiple—2, 3, 4 and so on—of the same irreducible value. Since the value of the charge, like the ratio of charge to mass, was thereby shown to be a constant, Millikan's experiment provided the conclusive evidence that all electrons were identical fundamental particles.

After completing the electron-charge work, Millikan returned to one of his earliest research subjects: the photoelectric effect. Einstein's quantum-based formula for the phenomenon had not yet been conclusively verified. Taking pains to avoid the mistakes of earlier experimenters, Millikan by 1915 had confirmed the validity of Einstein's equation in every detail, including the linearity of the relation between the maximum energy of the emitted electrons and the frequency of the incident light. In addition Millikan demonstrated that the slope of the line equaled the ratio of Planck's constant h to the elec-

tron charge, a result that supplied the best measure of h then available. In spite of his demonstration of the validity of Einstein's equation, Millikan did not believe he had confirmed the quantum theory of light on which it was based. Because of the overwhelming evidence for the wave nature of light he was sure, as many other contemporary physicists were, that the equation had to be based on a false (although obviously quite fruitful) hypothesis.

With the completion of the work on the electron charge and the photoelectric effect, the two investigations for which he was awarded his Nobel prize in 1923, Millikan came to the end of his first research program. He easily found new significant problems to explore. Not long after Millikan came to Chicago, Michelson made him responsible for supervising doctoral research. To find suitable dissertation topics Millikan read *Science Abstracts* regularly, and he became knowledgeable about the rich variety of questions arising in 20th-century physics. Before World War I intervened he managed some preliminary investigation of hot-spark spectra and of the "penetrating radiation" he later christened cosmic rays.

At Cal Tech in the 1920's, working with a hot spark between two electrodes in a vacuum, Millikan embarked with Ira S. Bowen, a graduate student, on a thorough study of the ultraviolet spectra of the lighter elements. By 1924 he and Bowen had extended the observable spectrum down to a wavelength of 1.36 angstrom units, helping to close the last gap between optical and X-ray frequencies. In the course of this work they discovered that, like hydrogen, atoms stripped of their valence (outer shell) electrons had spectra with doublets, or two closely spaced spectral lines. They also learned that certain of their ultraviolet doublets corresponded precisely to various energy levels associated with the X-ray spectra of the heavier elements. The German physicist Arnold Sommerfeld had accounted for these X-ray doublets by a relativistic analysis of atomic orbits. Millikan and Bowen suggested that Sommerfeld's relativistic approach might account for the doublets in the entire field of optical spectra.

They recognized that their results raised a serious difficulty for spectral theory. In Sommerfeld's relativistic scheme different X-ray doublet terms were assigned different quantum numbers, whereas in Niels Bohr's approach to optical spectra the doublets were accounted for by assuming different spatial orientations for orbits with the same quantum number. Since, according to Millikan and Bowen, X-ray and optical doublet terms were indistinguishable, one seemingly had to give up either Sommerfeld's relativistic explanation or the Bohr scheme of spectra. Millikan and Bowen could find no way out of the dilemma, but their forceful statement of it in 1924 contributed, as did that of the German physicist Alfred Landé, to the ultimate resolution of the difficulty through the postulation of electron spin by George E. Uhlenbeck and Samuel A. Goudsmit in 1925.

While pursuing the hot-spark work Millikan also resumed his investigation of penetrating radiation. In 1912 the Austrian-born physicist Victor F. Hess had demonstrated in a manned balloon flight up to 12,000 feet that atmospheric ionization increases with altitude, and on that evidence Hess argued that some type of radiation must be arriving from the heavens. Many scientists nonetheless continued to attribute atmospheric ionization to some terrestrial cause such as radioactivity. Millikan explored the issue by measuring ionization

first on top of Pike's Peak and then in the high atmosphere with electroscopes mounted on unmanned sounding balloons.

In the summer of 1925 he proposed to settle the question by measuring the variation of ionization with depth in Muir Lake and Lake Arrowhead in the mountains of California. The two lakes were snow-fed and were separated by many miles and by 6,675 feet of altitude (Muir Lake is the higher); each was likely to be free of both local terrestrial radioactive disturbances and any atmospheric peculiarities that might affect the level of ionization in the other. Millikan's electroscopic measurements showed that the intensity of ionization at any given depth below the surface of Lake Arrowhead was the same as the intensity six feet deeper than that in Muir Lake. Since the layer of atmosphere between the surfaces of the two lakes had the absorptive power of precisely six feet of water, the results decisively confirmed that the radiation was coming from the cosmos.

More penetrating than even the hardest gamma rays known, this cosmic radiation, in Millikan's belief, could not

consist of charged particles. Such particles would have to possess energies that were then thought to be impossibly high in order to penetrate, as the cosmic rays did, the combined air and water equivalent of six feet of lead. In Millikan's tentative opinion cosmic rays were likely to be photons, or quanta of electromagnetic radiation. The trajectories of incoming photons would not be affected by the earth's magnetic field; those of incoming charged particles would be affected by it, so that more of them would strike the earth at higher latitudes than would strike it at lower ones. In various experiments conducted in South America and at sea Millikan found no variation in intensity with latitude. In 1929, asked by reporters to comment on the notion that cosmic rays might be likened to charged particles, he remarked: "You might as well sensibly compare an elephant and a radish."

Other workers, notably the Dutch physicist Jacob Clay in 1927, had begun to detect signs of a latitude effect. Furthermore, in 1929 the German physicists Walther Bothe and Werner Kohlhörster found evidence, in coincidence experiments with two particle detectors, that at least a large fraction of cosmic

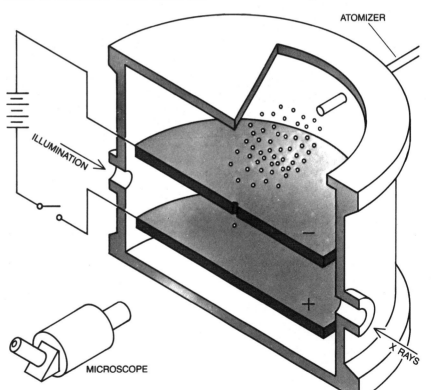

OIL-DROP EXPERIMENT that established the charge of the electron was conducted by Millikan with the simple equipment shown here very schematically. A droplet from an oil spray introduced into the upper chamber drifts through a pinhole into the space between two charged plates. The charge on the drop can be varied by ionizing the air around it with X rays; an electric field of known intensity can be established between the plates or can be turned off. The rate of rise or fall of the droplet is measured under various conditions by tracking it in the microscope. The various speeds turn out to be proportional to multiples of a single irreducible value: the electron's charge, which Millikan calculated to be $4.774 \pm .009 \times 10^{-10}$ electrostatic unit.

rays consisted of enormously energetic charged particles. Moreover, theorists had predicted that if cosmic rays consisted of charged particles, there should be an intensity difference between those arriving from the east and those arriving from the west. In 1932 Thomas H. Johnson, a young physicist at the Bartol Research Foundation in Philadelphia, detected a slight excess of particles coming from the west on Mount Washington in New Hampshire. Finally, also in 1932, following a worldwide survey at various altitudes, Arthur Holly Compton announced incontrovertible evidence of a latitude effect.

Millikan hotly contested the assertions that cosmic rays were charged particles, in particular Compton's report of a latitude effect. Millikan had repeated his own search for a latitude effect late in 1932 without success. The trouble, he later explained, was that in the longitudinal region of California the dip in cosmic-ray intensity began quite suddenly in the vicinity of Los Angeles and rapidly reached its maximum fall of some 7 percent less than two days' sail south of the city; in Millikan's initial search for the latitude effect his estimated error had been 6 percent. In most of his later searches he went to the north of Pasadena, where the rise in intensity was too small to detect easily. In 1932 he sent H. Victor Neher, a young collaborator at Cal Tech, on a voyage to the south, but Neher could not get his electroscope working until after he had passed the region of the dip.

By 1933, after Neher had found a latitude effect on the return voyage, Millikan was prepared to admit that at least some portion of cosmic radiation must consist of charged particles, but he insisted that any such rays must be secondary ones, produced in encounters between the incoming primaries and the nuclei of atoms in the atmosphere. He continued to maintain that the primaries, or at least some fraction of them, were photons. He clung tenaciously to one or another variation of that belief for 20 years in the face of overwhelming contrary evidence and the bulk of scientific opinion.

One can only speculate about why Millikan insisted on flying in the face of evidence and authority. Possibly it was because with the onset of age (in 1933 he was 65) he had grown scientifically dogmatic and self-centered; he infuriated other cosmic-ray researchers, notably Compton and Clay, who found him cloyingly self-congratulatory about his own work and unblushingly negligent about theirs. Possibly it was because he had committed himself openly and heavily, before the scientific and lay public alike, to the photon theory of cosmic rays. Possibly, too, it was because of his religious views. A Universalist in

sectarian affiliation and something akin to a deist in belief, Millikan saw the hand of the Creator beneficently at work in the origin (or at least in his particular theory of the origin) of cosmic rays that were photons.

Millikan's theory proceeded from the fact that no single coefficient of radiation absorption could account for the degree of ionization produced by cosmic rays as a function of depth in the atmosphere. Holding that the absorption curves could be broken down into three independent parts, he inferred that cosmic rays were clustered in three distinct energy bands, at 26, 110 and 220 million electron volts. He recognized that 26 MeV was just about the amount of energy equivalent to the "mass defect" of helium, that is, to the difference between the mass of four hydrogen atoms and the very slightly smaller mass of a helium atom formed by their fusion. Similarly, 110 MeV was close to the mass defects of oxygen and nitrogen, and 220 MeV was the mass defect of silicon.

Millikan concluded that the photons striking the earth must be generated when four atoms of hydrogen somehow fused to form helium, when 14 hydrogens fused to form nitrogen and 16 to form oxygen, and when 28 hydrogens fused to form silicon. Cosmic rays, then, were the "birth cries" of atoms, a Millikan phrase that achieved a good deal of currency. Drawing a remarkable religious inference from this theory, he proposed that such creation of the elements, which he called atom building, was proceeding continually, and that the attendant emission of energy arising from the mass defect was saving the universe from the heat death to which the second law of thermodynamics was alleged to have condemned it. The Creator, as Millikan explained the significance of his atom-building hypothesis, was "continually on His job."

Many physicists, even those who for a while preferred the photon interpretation, scoffed at Millikan's hypothesis, not least because of the kinetic difficulties that would have to be overcome for so many hydrogen atoms to collide simultaneously in space. Millikan wrote off the kinetic objections. After all, he declared, modern physics had moved away from a strictly mechanical view of nature. Nevertheless, by the mid-1930's he had to give up the atom-building hypothesis, mainly because experiments in his own laboratory as well as in others had demonstrated that the bulk of cosmic rays have energies far higher than can be accounted for by the mass defects of helium, nitrogen, oxygen and silicon.

Although he stopped worrying about whether or not the Creator was still on His job, Millikan remained interested in the origin of cosmic rays as a problem in science. In the late 1930's he proposed that cosmic rays originated in the spontaneous annihilation of atoms with an emission of high-energy photons. No more convincing than its predecessor, this hypothesis became untenable (as Millikan himself admitted a few years before his death) after the detection of the pi meson by C. F. Powell and others in 1947 made it clear that the primary cosmic radiation consists almost entirely of protons.

Still, after more than a decade of research, Millikan and his collaborators Neher and William H. Pickering had gathered considerable quantities of important data by measuring cosmic-ray intensities around the world, at sea level, in airplanes at high altitudes and with unmanned sounding balloons at even higher altitudes. In 1934, independently of Clay, Millikan detected the variation of the latitude effect with longitude because of the dissymmetry of the earth's magnetic field. Even the atom-building hypothesis yielded a significant divi-

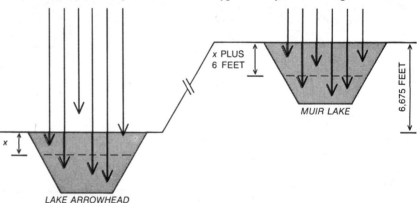

COSMIC-RAY EXPERIMENT devised by Millikan established that the radiation originates in the cosmos, not in the earth or in the lower atmosphere. He measured the ionization produced by cosmic rays in two lakes in California. The ionization intensity at any depth (x) in Lake Arrowhead was the same as the intensity six feet deeper than that ($x + 6$) in Muir Lake; the six feet compensated exactly for the 6,675 feet of air between the surfaces of the two lakes, showing, as Millikan wrote, "that the atmosphere between these two levels acted simply as an absorbing blanket and contributed not a bit to the intensity of the radiation found at the lower level."

THREE NOBEL PRIZEWINNERS in physics were photographed together in January of 1931, when A. A. Michelson (*left*), Albert Einstein and Millikan took part in a conference that was held at California Institute of Technology, which Millikan had headed since 1921.

AT CAL TECH IN 1930'S Millikan was photographed with physicists P. A. M. Dirac (*left*), who was visiting the institute, and J. Robert Oppenheimer, who had come to Cal Tech in 1929 as assistant professor of physics. Dirac won the Nobel prize in physics in 1933.

dend. Eager to obtain a direct measure of cosmic-ray energies, in 1931 Millikan set young Carl D. Anderson to studying the behavior of cosmic rays in a cloud chamber, and Anderson's experiments led to his detection of the positron the following year. As in the case of the positron, Millikan's data had all been obtained in pursuit of mistaken hypotheses, yet the data were no less valuable and the effort was no less striking for the wrongheadedness of their rationale.

Millikan's dispute with Compton, like his "birth cries of atoms" hypothesis, received considerable attention in the press. Millikan decried sensationalism, but the fact of the matter was that he was an avid publicist, one of the best in the scientific community. Articles by or about him were published frequently in both the local and the national press. Radio listeners heard him discourse in his Yankee-sounding twang on the New Deal or the new physics. Irreverent Cal Tech faculty members spoke of the "milli-kan": one thousandth of a unit of publicity.

As a public figure Millikan was most popular and respected in the 1920's, when he won his Nobel prize. (In 1927 he made the cover of *Time*, which noted that he had the face of a "witty and successful banker.") The decade of "normalcy" was also a period of unprecedented interest in science: in science as such, as a source of marvelous technology and also, as the Scopes trial made clear, as a challenge to religious authority. The Reverend Harry Emerson Fosdick was moved to say: "When a prominent scientist comes out strongly for religion, all the churches thank Heaven and take courage as though it were the highest possible compliment to God to have Eddington believe in Him." Millikan spoke out strongly for religion, for moral and spiritual values and for a faith tempered by the open-minded tolerance of the scientific spirit. He also celebrated science as an ally of the laissez faire economic ideas prevalent in the 1920's. The resolution of want, he believed, lay not in governmental economic intervention but in more abundant production by a more abundant industry. "No efforts toward social readjustments or toward the redistribution of wealth," he asserted, "have one thousandth as large a chance of contributing to human well-being as have the efforts of the physicist, the chemist and the biologist toward the better understanding and better control of nature."

Millikan's declarations drew criticism, notably from humanists who chastized him for seeming to argue that morality progressed with the progress of research. While modern man knew more than Socrates, they pointed out, he was neither wiser nor more decent. Industry wrote checks on the intellectual bank account of the sciences, but the sciences did not in turn check the rapacious industrialist. And if science had beneficently enlarged peaceful man's mastery over nature, the humanists stressed, it had also multiplied warlike man's power to kill and destroy. During the Depression the attacks on science became more strident. In response Millikan emerged not only as an exponent of science but also as a staunch defender of it. A conservative Republican, he maintained in the 1930's that investing in basic research was superior to Federal meddling in the economy. The further advancement of science, he argued, would lead to new technology, new industries and ultimately new jobs.

For the sake of economic growth following World War I and of economic recovery in the 1930's, Millikan backed Federal aid to academic science, but he also harbored doubts about it. On the one hand, it was clear that public funds would enlarge the opportunities for investigation and training; on the other, he worried, funds from the Government would inevitably "politicize" science. He was usually vague about whether this politicizing would subject the topics and results of research to intellectual censorship, programmatic control or something else. Whatever he meant specifically, in general he feared Federal interference with the autonomy of academic investigators and their institutions. To avoid that he insisted it was essential to insulate the control of Federal research funds from political manipulation, which in his lexicon meant from politically responsible officials, elected or appointed. To that end Millikan consistently proposed to vest control over public money for academic science directly or indirectly in the National Academy of Sciences. Just as consistently, public officials repeatedly balked at turning over any such public function to a private agency.

Whereas before 1940 most of Millikan's scientific contemporaries had agreed with his view of Federal funding of research, he later found himself increasingly isolated in this respect. After World War II most leading American scientists endorsed the establishment of the National Science Foundation, a governmental body designed with safeguards against crude political control. In spite of the safeguards, Millikan was willing to go along only with a foundation whose membership was controlled by the Academy. He adamantly opposed the establishment of what he called "a new large Federal agency politically responsible and politically controlled." He was appalled that "the great

RECORDING ELECTROSCOPE displayed by Millikan in 1935 was a late model of the kind of equipment with which he had for years been studying ionization caused by cosmic rays.

majority of scientists" had evidently lined up behind "pork-barrel science bills." His distress stemmed not only from his old fear that a system of Federal aid to academic institutions would lead to "the ultimate control of education by the party in power" but also from his concern about scientists getting into the business of pleading for public support from Washington. "Despite all our protestations of holiness in motive," he warned, the public would come to see scientists as just another interest group, like the merchant marine, commercial airlines or agriculture.

Most American scientists, however, endorsed the creation of the vast postwar Federal apparatus for aid to academic research, including not only the National Science Foundation but also the Atomic Energy Commission, the expanded National Institutes of Health and the Office of Naval Research. The dissent of Millikan and his few allies was overwhelmed by the flow of Federal funds into university science, the rising affluence of the scientific community and the subsequent American command of world science. Millikan had maintained that private resources could continue to suffice for academic research as they had before 1940. Few scientists agreed with him, and doubtless he was wrong on that score. On the other hand, few thoughtful scientists or academic administrators nowadays would find fault with Millikan for his fears. Even if the Federal Government is far from having seized totalitarian control of education and research, it has begun to exercise a degree of control—some would say excessive interference—in American universities that many academic scientists find uncomfortable, to say the least. And if the public credibility and authority of the scientific community have diminished in recent years, one reason may have been the rising suspicion that scientists are concerned at least as much with self-advancement as with the advancement of knowledge.

In his autobiography Millikan said little, apart from propounding some general strictures against Federal aid to education, about his role in the historical attempts to achieve Government funding of academic science. He devoted

BALLOON EXPERIMENT designed to measure cosmic-ray intensities in the stratosphere brought Millikan (*left*) to Fort Sam Houston in Texas in August of 1935. He did valuable research on cosmic rays in spite of the idiosyncratic nature of his hypotheses of their origin.

only a few pages to the years of labor with cosmic rays. By 1950 he obviously found both subjects awkward to recall, and in the case of cosmic rays perhaps even painful. The autobiography may have dwelt on the earlier chapters of his life not simply because (as is often the case) they loomed larger and sharper in memory but also because they were chapters of unmitigated success.

Yet Millikan could have taken a certain pride in the later chapters of comparative failure. If it is true that he had fallen increasingly out of step on the issue of Federal patronage, it is also true that in his opposition he had raised questions that would return to trouble the scientific community. And if he was on the losing side in the debate over the nature of cosmic rays, in searching for their origin he had tackled a problem of no small difficulty. In 1950 Millikan at-tached a note at the top of his last paper on the atom-annihilation hypothe-sis: "New evidence has appeared since this was written which is unfavorable to this hypothesis, but the experimental data herein contained is valid. The actual origin of the cosmic rays is still today an unsolved mystery." The mystery re-mains unsolved.

12

Prematurity and Uniqueness in Scientific Discovery

by Gunther S. Stent
December 1972

A molecular geneticist reflects on two general historical questions: (1) What does it mean to say a discovery is "ahead of its time"? (2) Are scientific creations any less unique than artistic creations?

The fantastically rapid progress of molecular genetics in the past 25 years now obliges merely middle-aged participants in its early development to look back on their early work from a depth of historical perspective that for scientific specialties flowering in earlier times came only after all the witnesses of the first blossoming were long dead. It is as if the late-18th-century colleagues of Joseph Priestley and Antoine Lavoisier had still been active in chemical research and teaching in the 1930's, after atomic structure and the nature of the chemical bond had been revealed. This somewhat depressing personal vantage provides a singular opportunity to assay the evolution of a scientific field. In reflecting on the history of molecular genetics from the viewpoint of my own experience I have found that two of its most famous incidents—Oswald Avery's identification of DNA as the active principle in bacterial transformation and hence as genetic material, and James Watson and Francis Crick's discovery of the DNA double helix—illuminate two general problems of cultural history. The case of Avery throws light on the question of whether it is meaningful or merely tautologous to say that a discovery is "ahead of its time," or premature. And the case of Watson and Crick can be used, and in fact has been used, to discuss the question of whether there is anything unique in a scientific discovery, in view of the likelihood that if Dr. *A* had not discovered Fact *X* today, Dr. *B* would have discovered it tomorrow.

Five years ago I published a brief retrospective essay on molecular genetics, with particular emphasis on its origins. In that historical account I mentioned neither Avery's name nor DNA-mediated bacterial transformation. My essay elicited a letter to the editor by a microbiologist, who complained: "It is a sad and surprising omission that... Stent makes no mention of the definitive proof of DNA as the basic hereditary substance by O. T. Avery, C. M. MacLeod and Maclyn McCarty. The growth of [molecular genetics] rests upon this experimental proof.... I am old enough to remember the excitement and enthusiasm induced by the publication of the paper by Avery, MacLeod and McCarty. Avery, an effective bacteriologist, was a quiet, self-effacing, non-disputatious gentleman. These characteristics of personality should not [cause] the general scientific public... to let his name go unrecognized."

I was taken aback by this letter and replied that I should indeed have mentioned Avery's 1944 proof that DNA is the hereditary substance. I went on to say, however, that in my opinion it is not true that the growth of molecular genetics rests on Avery's proof. For many years that proof actually had little impact on geneticists. The reason for the delay was not that Avery's work was unknown to or mistrusted by geneticists but that it was "premature."

My *prima facie* reason for saying Avery's discovery was premature is that it was not appreciated in its day. By lack of appreciation I do not mean that Avery's discovery went unnoticed, or even that it was not considered important. What I do mean is that geneticists did not seem to be able to do much with it or build on it. That is, in its day Avery's discovery had virtually no effect on the general discourse of genetics.

This statement can be readily supported by an examination of the scientific literature. For example, a convincing demonstration of the lack of appreciation of Avery's discovery is provided by the 1950 golden jubilee of genetics symposium "Genetics in the 20th Century." In the proceedings of that symposium some of the most eminent geneticists published essays that surveyed the progress of the first 50 years of genetics and assessed its status at that time. Only one of the 26 essayists saw fit to make more than a passing reference to Avery's discovery, then six years old. He was a colleague of Avery's at the Rockefeller Institute, and he expressed some doubt that the active transforming principle was really pure DNA. The then leading philosopher of the gene, H. J. Muller of Indiana University, contributed an essay on the nature of the gene that mentions neither Avery nor DNA.

So why was Avery's discovery not appreciated in its day? Because it was "premature." But is this really an explanation or is it merely an empty tautology? In other words, is there a way of providing a criterion of the prematurity of a discovery other than its failure to make an impact? Yes, there is such a criterion: A discovery is premature if its implications cannot be connected by a series of simple logical steps to canonical, or generally accepted, knowledge.

Why could Avery's discovery not be connected with canonical knowledge? Ever since DNA had been discovered in the cell nucleus by Friedrich Miescher in 1869 it had been suspected of exerting some function in hereditary processes. This suspicion became stronger in the 1920's, when it was found that DNA is a major component of the chromosomes. The then current view of the molecular nature of DNA, however, made it well-nigh inconceivable that DNA could be the carrier of hereditary information. First, until well into the 1930's DNA was generally thought to be merely a tetranucleotide composed of one unit each of adenylic, guanylic, thymidylic and cytidylic acids. Second, even when it was finally realized by the early 1940's that the molecular weight of DNA is actually much higher than the tetranucleotide hypothesis required, it was still widely believed the tetranucleotide was the basic repeating unit of the large DNA polymer in which the four units mentioned recur in regular sequence. DNA was therefore viewed as a uniform macromolecule that, like other monotonous polymers such as starch or cellulose, is always the same, no matter what its biological source. The ubiquitous presence of DNA in the chromosomes was therefore generally explained in purely physiological or structural terms. It was usually to the chromosomal protein that the informational role of the genes had been assigned, since the great differences in the specificity of structure that exist between various proteins in the same or-

PICASSO'S "LES DESMOISELLES D'AVIGNON," painted in Paris in 1907, is often cited by art historians as the first major Cubist painting and a milestone in the development of modern art. It is reproduced here as an archetype of the proposition that works of artistic creation are unique (in the sense that if Picasso had not existed, it would never have been painted), whereas works of scientific creation are inevitable (in the sense that if Dr. A had not discovered Fact X today, Dr. B would discover it tomorrow). The validity of the proposition is disputed by the author. The painting is in the collection of the Museum of Modern Art in New York.

ganism, or between similar proteins in different organisms, had been appreciated since the beginning of the century. The conceptual difficulty of assigning the genetic role to DNA had not escaped Avery. In the conclusion of his paper he stated: "If the results of the present study of the transforming principle are confirmed, then nucleic acids must be regarded as possessing biological specificity the chemical basis of which is as yet undetermined."

By 1950, however, the tetranucleotide hypothesis had been overthrown, thanks largely to the work of Erwin Chargaff of the Columbia University College of Physicians and Surgeons. He showed that, contrary to the demands of that hypothesis, the four nucleotides are not necessarily present in DNA in equal proportions. He found, furthermore, that the exact nucleotide composition of DNA differs according to its biological source, suggesting that DNA might not be a monotonous polymer after all. And so when two years later, in 1952, Alfred Hershey and Martha Chase of the Carnegie Institution's laboratory in Cold Spring Harbor, N.Y., showed that on infection of the host bacterium by a bacterial virus at least 80 percent of the viral DNA enters the cell and at least 80 percent of the viral protein remains outside, it was possible to connect their conclusion that DNA is the genetic material with canonical knowledge. Avery's "as yet undetermined chemical basis of the biological specificity of nucleic acids" could now be seen as the precise sequence of the four nucleotides along the polynucleotide chain. The general impact of the Hershey-Chase experiment was immediate and dramatic. DNA was suddenly in and protein was out, as far as thinking about the nature of the gene was concerned. Within a few months there arose the first speculations about the genetic code, and Watson and Crick were inspired to set out to discover the structure of DNA.

Of course, Avery's discovery is only one of many premature discoveries in the history of science. I have presented it here for consideration mainly because of my own failure to appreciate it when I joined Max Delbrück's bacterial virus group at the California Institute of Technology in 1948. Since then I have often wondered what my later career would have been like if I had only been astute enough to appreciate Avery's discovery and infer from it four years before Hershey and Chase that DNA must

also be the genetic material of our own experimental organism.

Probably the most famous case of prematurity in the history of biology is associated with the name of Gregor Mendel, whose discovery of the gene in 1865 had to wait 35 years before it was "rediscovered" at the turn of the century. Mendel's discovery made no immediate impact, it can be argued, because the concept of discrete hereditary units could not be connected with canonical knowledge of anatomy and physiology in the middle of the 19th century. Furthermore, the statistical methodology by means of which Mendel interpreted the results of his pea-breeding experiments was entirely foreign to the way of thinking of contemporary biologists. By the end of the 19th century, however, chromosomes and the chromosome-dividing processes of mitosis and meiosis had been discovered and Mendel's results could now be accounted for in terms of structures visible in the microscope. Moreover, by then the application of statistics to biology had become commonplace. Nonetheless, in some respects Avery's discovery is a more dramatic example of prematurity than Mendel's. Whereas Mendel's discovery seems hardly to have been mentioned by anyone until its rediscovery, Avery's discovery was widely discussed and yet it could not be appreciated for eight years.

Cases of delayed appreciation of a discovery exist also in the physical sciences. One example (as well as an explanation of its circumstances in terms of the concept to which I refer here as prematurity) has been provided by Michael Polanyi on the basis of his own experience. In the years 1914–1916 Polanyi published a theory of the adsorption of gases on solids which assumed that the force attracting a gas molecule to a solid surface depends only on the position of the molecule, and not on the presence of other molecules, in the force field. In spite of the fact that Polanyi was able to provide strong experimental evidence in favor of his theory, it was generally rejected. Not only was the theory rejected, it was also considered so ridiculous by the leading authorities of the time that Polanyi believes continued defense of his theory would have ended his professional career if he had not managed to publish work on more palatable ideas. The reason for the general rejection of Polanyi's adsorption theory was that at the very time he put it forward the role of elec-

trical forces in the architecture of matter had just been discovered. Hence there seemed to be no doubt that the adsorption of gases must also involve an electrical attraction between the gas molecules and the solid surface. That point of view, however, was irreconcilable with Polanyi's basic assumption of the mutual independence of individual gas molecules in the adsorption process. It was only in the 1930's, after a new theory of cohesive molecular forces based on quantum-mechanical resonance rather than on electrostatic attraction had been developed, that it became conceivable gas molecules could behave in the way Polanyi's experiments indicated they were actually behaving. Meanwhile Polanyi's theory had been consigned so authoritatively to the ashcan of crackpot ideas that it was rediscovered only in the 1950's.

Still, can the notion of prematurity be said to be a useful historical concept? First of all, is prematurity the only possible explanation for the lack of contemporary appreciation of a discovery? Evidently not. For example, my microbiologist critic suggested that it was the "quiet, self-effacing, non-disputatious" personality of Avery that was the cause of the failure of his contribution to be recognized. Furthermore, in an essay on the history of DNA research Chargaff supports the idea that personal modesty and aversion to self-advertisement account for the lack of contemporary scientific appreciation. He attributes the 75-year lag between Miescher's discovery of DNA and the general appreciation of its importance to Miescher's being "one of the quiet in the land," who lived when "the giant publicity machines, which today accompany even the smallest move on the chess-board of nature with enormous fanfares, were not yet in place." Indeed, the 35-year hiatus in the appreciation of Mendel's discovery is often attributed to Mendel's having been a modest monk living in an out-of-the-way Moravian monastery. Hence the notion of prematurity provides an alternative to the invocation—in my opinion an inappropriate one for the instances mentioned here—of the lack of publicity as an explanation for delayed appreciation.

More important, does the prematurity concept pertain only to retrospective judgments made with the wisdom of hindsight? No, I think it can be used also to judge the present. Some recent discoveries are still premature at this very time. One example of here-and-now prematurity is the alleged finding that experiential information received by an animal can be stored in nucleic acids or other macromolecules.

Some 10 years ago there began to appear reports by experimental psychologists purporting to have shown that the engram, or memory trace, of a task learned by a trained animal can be transferred to a naïve animal by injecting or feeding the recipient with an extract made from the tissues of the donor. At that time the central lesson of molecular genetics—that nucleic acids and proteins are informational macromolecules—had just gained wide currency, and the facile equation of nervous information with genetic information soon led to the proposal that macromolecules—DNA, RNA or protein—store memory. As it happens, the experiments on which the macromolecular theory of memory is based have been difficult to repeat, and the results claimed for them may indeed not be true at all. It is nonetheless significant that few neurophysiologists have even bothered to check these experiments, even though it is common knowledge that the possibility of chemical memory transfer would constitute a fact of capital importance. The lack of interest of neurophysiologists in the macromolecular theory of memory can be accounted for by recognizing that the theory, whether true or false, is clearly premature. There is no chain of reasonable inferences by means of which our present, albeit highly imperfect, view of the functional organization of the brain can be reconciled with the possibility of its acquiring, storing and retrieving nervous information by encoding such information in molecules of nucleic acid or protein. Accordingly for the community of neurophysiologists there is no point in devoting time to checking on experiments whose results, even if they were true as alleged, could not be connected with canonical knowledge.

The concept of here-and-now prematurity can be applied also to the troublesome subject of ESP, or extrasensory perception. In the summer of 1948 I happened to hear a heated argument at Cold Spring Harbor between two future mandarins of molecular biology, Salvador Luria of Indiana University and R. E. Roberts of the Carnegie Institution's laboratory in Washington. Roberts was then interested in ESP, and he felt it had not been given fair consideration by the scientific community. As I recall, he thought that one might be able to set up experiments with molecular beams that could provide more definitive data on the possibility of mind-induced departures from random distributions than J. B. Rhine's then much discussed card-guessing procedures. Luria declared that not only was he not interested in Roberts' proposed experiments but also in his opinion it was unworthy of anyone claiming to be a scientist even to discuss such rubbish. How could an intelligent fellow such as Roberts entertain the possibility of phenomena totally irreconcilable with the most elementary physical laws? Moreover, a phenomenon that is manifest only to specially endowed subjects, as claimed by "parapsychologists" to be the case for ESP, is outside the proper realm of science, which must deal with phenomena accessible to every observer. Roberts replied that far from him being unscientific, it was Luria whose bigoted attitude toward the unknown was unworthy of a true scientist. The fact that not everyone has ESP only means that it is an elusive phenomenon, similar to musical genius. And just because a phenomenon cannot be reconciled with what we now know, we need not shut our eyes to it. On the contrary, it is the duty of the scientist to try to devise experiments designed to probe its truth or falsity.

It seemed to me then that both Luria and Roberts were right, and in the intervening years I often thought about this puzzling disagreement, unable to resolve it in my own mind. Finally six years ago I read a review of a book on ESP by my Berkeley colleague C. West Churchman, and I began to see my way toward a resolution. Churchman stated that there are three different possible scientific approaches to ESP. The first of these is that the truth or falsity of ESP, like the truth or falsity of the existence of God or of the immortality of the soul, is totally independent of either the methods or the findings of empirical science. Thus the problem of ESP is defined out of existence. I imagine that this was more or less Luria's position.

Churchman's second approach is to reformulate the ESP phenomenon in terms of currently acceptable scientific notions, such as unconscious perception or conscious fraud. Hence, rather than defining ESP out of existence, it is trivialized. The second approach probably would have been acceptable to Luria too, but not to Roberts.

The third approach is to take the proposition of ESP literally and to attempt to examine in all seriousness the

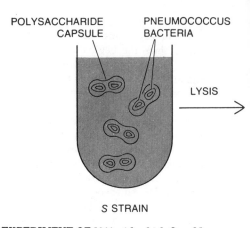

POLYSACCHARIDE CAPSULE PNEUMOCOCCUS BACTERIA

LYSIS

S STRAIN

EXPERIMENT OF 1944 with which Oswald Avery correctly identified the chemical nature of the genetic material is regarded by the author as a classic example of a premature scientific discovery. The virulent normal, or S-type, pneumococcus, a bacteri-

evidence for its validity. That was more or less Roberts' position. As Churchman points out, however, this approach is not likely to lead to satisfactory results. Parapsychologists can maintain with some justice that the existence of ESP has already been proved to the hilt, since no other set of hypotheses in psychology has received the degree of critical scrutiny that has been given to ESP experiments. Moreover, many other phenomena have been accepted on much less statistical evidence than what is offered for ESP. The reason Churchman advances for the futility of a strictly evidential approach to ESP is that in the absence of a hypothesis of how ESP could work it is not possible to decide whether any set of relevant observations can be accounted for only by ESP to the exclusion of alternative explanations.

After reading Churchman's review I realized that Roberts would have been ill-advised to proceed with his ESP experiments, not because, as Luria had claimed, they would not be "science" but because any positive evidence he might have found in favor of ESP would have been, and would still be, premature. That is, until it is possible to connect ESP with canonical knowledge of, say, electromagnetic radiation and neurophysiology no demonstration of its occurrence could be appreciated.

Is the lack of appreciation of premature discoveries merely attributable to the intellectual shortcoming or innate conservatism of scientists who, if they were only more perceptive or more open-minded, would give immediate recognition to any well-documented scientific proposition? Polanyi is not of that opinion. Reflecting on the cruel fate of

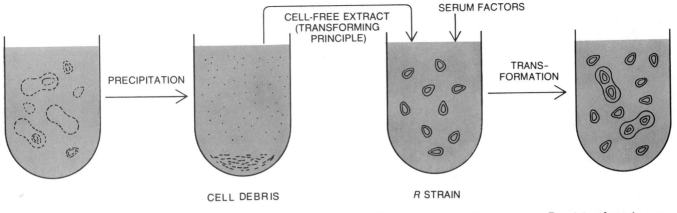

CELL DEBRIS   R STRAIN

um that causes pneumonia in mammals, is enclosed in a smooth (hence *S*) polysaccharide capsule that protects the bacterium from the ordinary defense mechanisms of the infected animal. The avirulent mutant, or *R*-type (*R* for rough), strain has lost the genetic capacity to form this protective capsule and hence is comparatively harmless. When a "transforming principle" extracted from normal

S donor bacteria was added to mutant *R* recipient bacteria, some of the mutants were found to regain the genetic capacity to form the capsule and thus were transformed back into the normal, virulent *S* type. Avery purified the transforming principle and succeeded in showing that it is DNA. The significance of Avery's discovery was not appreciated by molecular geneticists until 1952.

his theory half a century after first advancing it, he declared: "This miscarriage of the scientific method could not have been avoided.... There must be at all times a predominantly accepted scientific view of the nature of things, in the light of which research is jointly conducted by members of the community of scientists. A strong presumption that any evidence which contradicts this view is invalid must prevail. Such evidence has to be disregarded, even if it cannot be accounted for, in the hope that it will eventually turn out to be false or irrelevant."

That is a view of the operation of science rather different from the one commonly held, under which acceptance of authority is seen as something to be avoided at all costs. The good scientist is seen as an unprejudiced man with an open mind who is ready to embrace any new idea supported by the facts. The history of science shows, however, that its practitioners do not appear to act according to that popular view.

Five years ago Chargaff wrote one of the many reviews of *The Double Helix*, Watson's autobiographical account of his and Crick's discovery of the structure of DNA. In his review Chargaff observes that scientific autobiography is "a most awkward literary genre." Most such works, he says, "give the impression of having been written for the remainder tables of bookstores, reaching them almost before they are published." The reasons for this, according to Chargaff, are not far to seek: scientists "lead monotonous and uneventful lives and ... besides often do not know how to write." Moreover, "there may also be profound-

er reasons for the general triteness of scientific autobiographies. *Timon of Athens* could not have been written, 'Les Desmoiselles d'Avignon' not have been painted, had Shakespeare and Picasso not existed. But of how many scientific achievements can this be claimed? One could almost say that, with very few exceptions, it is not the men that make science, it is science that makes the men. What *A* does today, *B* or *C* or *D* could surely do tomorrow."

On reading this passage, I found myself in full agreement on the general lack of literary skills among men of science. I was surprised, however, to find an eminent scientist embracing historicism (the theory championed by Hegel and Marx holding that history is determined by immutable forces rather than by human agency) as an explanation for the evolution of science while at the same time professing belief in the libertarian "great man" view of history for the evolution of art. Since it had not occurred to me that anyone could hold such contradictory, and to me obviously false, views concerning these two most important domains of human creation, I began to ask scientific friends and colleagues whether they too, by any chance, thought there was an important qualitative difference between the achievements of art and of science, namely that the former are unique and the latter inevitable. To my even greater surprise, I found that most of them seemed to agree with Chargaff. Yes, they said, it is quite true that we would not have had *Timon of Athens* or "Les Desmoiselles d'Avignon" if Shakespeare and Picasso had not existed, but if Watson and Crick had not existed, we would

have had the DNA double helix anyway. Therefore, contrary to my first impression, it does not seem to be all that obvious that this proposition has little philosophical or historical merit. Hence I shall now attempt to show that there is no such profound difference between the arts and sciences in regard to the uniqueness of their creations.

Before discussing the proposition of differential uniqueness of creation it is necessary to make an explicit statement of the meaning of "art" and of "science." My understanding of these terms is based on the view that both the arts and the sciences are activities that endeavor to discover and communicate truths about the world. The domain to which the artist addresses himself is the inner, subjective world of the emotions. Artistic statements therefore pertain mainly to relations between private events of affective significance. The domain of the scientist, in contrast, is the outer, objective world of physical phenomena. Scientific statements therefore pertain mainly to relations between or among public events. Thus the transmission of information and the perception of meaning in that information constitute the central content of both the arts and the sciences. A creative act on the part of either an artist or a scientist would mean his formulation of a new meaningful statement about the world, an addition to the accumulated capital of what is sometimes called "our cultural heritage." Let us therefore examine the proposition that only Shakespeare could have formulated the semantic structures represented by *Timon*, whereas people other than Watson and Crick might have made the communication

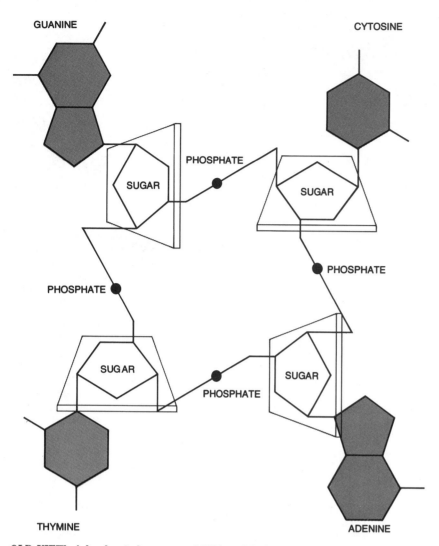

GUANINE

CYTOSINE

PHOSPHATE

SUGAR

SUGAR

PHOSPHATE

PHOSPHATE

SUGAR

SUGAR

PHOSPHATE

THYMINE

ADENINE

OLD VIEW of the chemical structure of DNA, widely held until well into the 1930's, saw the molecule as being merely a tetranucleotide composed of one unit each of adenylic, guanylic, thymidylic and cytidylic acids. This hypothesis demanded that the molecular weight of DNA be little more than 1,000 and that the four nucleotide bases (adenine, guanine, thymine and cytosine) occur in exactly equal proportions. Even when it was finally realized in the 1940's that the molecular weight of DNA is much higher (in the millions or billions), it was still widely believed that the tetranucleotide was the basic repeating unit of the large DNA polymer. The mistaken belief in this uniform macromolecular structure proved to be an obstacle to the eventual acceptance of the idea that DNA is the genetic material.

represented by their paper, "A Structure for Deoxyribonucleic Acid," published in *Nature* in the spring of 1953.

First, it is evident that the exact word sequence that Watson and Crick published in *Nature* would not have been written if the authors had not existed, any more than the exact word sequence of *Timon* would have been written without Shakespeare, at least not until the fabulous monkey typists complete their random work at the British Museum. And so both creations are from that point of view unique. We are not really concerned, however, with the exact word sequence. We are concerned with the content. Thus we admit that people other than Watson and Crick would

eventually have described a satisfactory molecular structure for DNA. But then the character of *Timon* and the story of his trials and tribulations not only might have been written without Shakespeare but also were written without him. Shakespeare merely reworked the story of *Timon* he had read in William Painter's collection of classic tales, *The Palace of Pleasure,* published 40 years earlier, and Painter in turn had used as his sources Plutarch and Lucian. But then we do not really care about Timon's story; what counts are the deep insights into human emotions that Shakespeare provides in his play. He shows us here how a man may make his response to the injuries of life, how he may turn

from lighthearted benevolence to passionate hatred toward his fellow men. Can one be sure, however, that *Timon* is unique from this bare-bones standpoint of the work's artistic essence? No, because who is to say that if Shakespeare had not existed no other dramatist would have provided for us the same insights? Another dramatist would surely have used an entirely different story (as Shakespeare himself did in his much more successful *King Lear*) to treat the same theme and he might have succeeded in pulling it off. The reason no one seems to have done it since is that Shakespeare had already done it in 1607, just as no one discovered the structure of DNA after Watson and Crick had already discovered it in 1953.

Hence we are finally reduced to asserting that *Timon* is uniquely Shakespeare's, because no other dramatist, although he might have brought us more or less the same insights, would have done it in quite the same exquisite way as Shakespeare. But here we must not shortchange Watson and Crick and take for granted that those other people who eventually would have found the structure of DNA would have found it in just the same way and produced the same revolutionary effect on contemporary biology. On the basis of my acquaintance with the personalities then engaged in trying to uncover the structure of DNA, I believe that if Watson and Crick had not existed, the insights they provided in one single package would have come out much more gradually over a period of many months or years. Dr. *B* might have seen that DNA is a double-strand helix, and Dr. *C* might later have recognized the hydrogen bonding between the strands. Dr. *D* later yet might have proposed a complementary purine-pyrimidine bonding, with Dr. *E* in a subsequent paper proposing the specific adenine-thymine and guanine-cytosine nucleotide pairs. Finally, we might have had to wait for Dr. *G* to propose the replication mechanism of DNA based on the complementary nature of the two strands. All the while Drs. *H, I, J, K* and *L* would have been confusing the issue by publishing incorrect structures and proposals. Thus I fully agree with the judgment offered by Sir Peter Medawar in his review of *The Double Helix:* "The great thing about [Watson and Crick's] discovery was its completeness, its air of finality. If Watson and Crick had been seen groping toward an answer, if they had published a partly right solution and had been obliged to follow it up with corrections and glosses, some of them made by other people; if the solution had come out piecemeal instead of in a blaze of understanding; then it would still have been a great episode in biological history; but something more in the common run of things; something splendidly well done, but not in the grand romantic manner."

Why is it that so many scientists apparently fail to see that it can be said of both art and science that whereas "what A does today, B or C or D could surely do tomorrow," B or C or D might nevertheless not do it as well as A, in the same "grand romantic manner." I think a variety of reasons can be put forward to account for this strange myopia. The first of them is simply that most scientists are not familiar with the working methods of artists. They tend to picture the artist's act of creation in the terms of Hollywood: Cornel Wilde in the role of the one and only Frédéric Chopin gazing fondly at Merle Oberon as his muse and mistress George Sand and then sitting down at the Pleyel pianoforte to compose his "Preludes." As scientists know full well, science is done quite differently: Dozens of stereotyped and ambitious researchers are slaving away in as many identical laboratories, all trying to make similar discoveries, all using more or less the same knowledge and techniques, some of them succeeding and some not. Artists, on the other hand, tend to conceive of the scientific act of creation in equally unrealistic terms: Paul Muni in the role of the one and only Louis Pasteur, who while burning the midnight oil in his laboratory has the inspiration to take some bottles from the shelf, mix their contents and thus discover the vaccine for rabies. Artists, in turn, know that art is done quite differently: Dozens of stereotyped and ambitious writers, painters and composers are slaving away in as many identical garrets, all trying to produce similar works, all using more or less the same knowledge and techniques, some succeeding and some not.

A second reason is that the belief in the inevitability of scientific discoveries appears to derive support from the often-told tales of famous cases in the history of science where the same discovery was made independently two or more times by different people. For instance, the independent invention of the calculus by Leibniz and Newton or the independent recognition of the role of natural selection in evolution by Wallace and Darwin. As the study of such "multiple discoveries" by Robert Merton of Columbia University has shown, however, on detailed examination they are rarely, if ever, identical. The reason they are said to be multiple is simply that in spite of their differences one can recognize a semantic overlap between them that is transformable into a congruent set of ideas.

The third, and somewhat more profound, reason is that whereas the cumulative character of scientific creation is at once apparent to every scientist, the similarly cumulative character of artistic creation is not. For instance, it is obvious that no present-day working geneticist has any need to read the original papers of Mendel, because they have been completely superseded by the work of the past century. Mendel's papers contain no useful information that cannot be better obtained from any modern textbook or the current genetical literature. In contrast, the modern writer, composer or painter still needs to read, listen or look at the original works of Shakespeare, Bach or Leonardo, which, so it is thought, have not been superseded at all. In spite of the seeming truth of this proposition, it must be said that art is no less cumulative than science, in that artists no more work in a traditionless vacuum than scientists do. Artists also build on the work of their predecessors; they start with and later improve on the styles and insights that have been handed down to them from their teachers, just as scientists do. To stay with our main example, Shakespeare's *Timon* has its roots in the works of Aeschylus, Sophocles and Euripides. It was those authors of Greek antiquity who discovered tragedy as a vehicle for communicating deep insights into affects, and Shakespeare, drawing on many earlier sources, finally developed that Greek discovery to its ultimate height. To some limited extent, therefore, the plays of the Greek dramatists have been superseded by Shakespeare's. Why, then, have Shakespeare's plays not been superseded by the work of later, lesser dramatists?

Here we finally do encounter an important difference between the creations of art and of science, namely the feasibility of paraphrase. The semantic content of an artistic work—a play, a cantata or a painting—is critically dependent on the exact manner of its realization; that is, the greater an artistic work is, the more likely it is that any omissions or changes from the original detract from its content. In other words, to paraphrase a great work of art—for instance

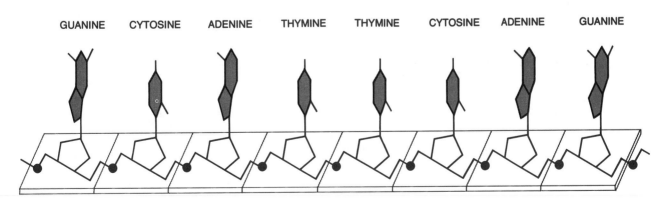

PRESENT VIEW of the chemical structure of DNA sees the molecule as a long chain in which the four nucleotide bases can be arranged in any arbitrary order. Although the proportion of adenine is always equal to that of thymine and the proportion of guanine is always equal to that of cytosine, the ratio of adenine-thymine to guanine-cytosine can vary over a large range, depending on the biological source of the DNA. With the elaboration of this single-strand structure it became possible to envision that genetic information is encoded in the DNA molecule as a specific sequence of the four nucleotide bases (*see illustration on next page*).

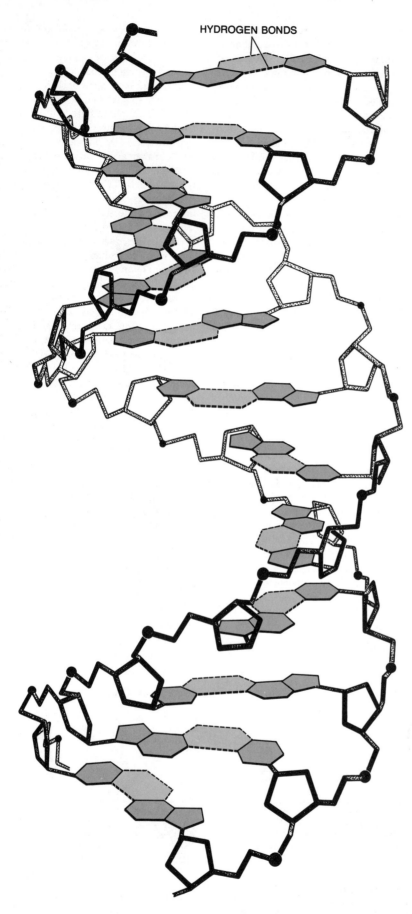

HYDROGEN BONDS

WATSON-CRICK MODEL of the structure of DNA, the discovery of which was announced in 1953, can now be described adequately as a double-strand self-complementary helix.

to rewrite *Timon*—without loss of artistic quality requires a genius equal to the genius of the original creator. Such a successful paraphrase would, in fact, constitute a great work of art in its own right. The semantic content of a great scientific paper, on the other hand, although its impact at the time of publication may also be critically dependent on the exact manner in which it is presented, can later be paraphrased without serious loss of semantic content by lesser scientists. Thus the simple statement "DNA is a double-strand, self-complementary helix" now suffices to communicate the essence of Watson and Crick's great discovery, whereas "A man responds to the injuries of life by turning from lighthearted benevolence to passionate hatred toward his fellow men" is merely a platitude and not a paraphrase of *Timon*. It took the writing of *King Lear* to paraphrase (and improve on) *Timon*, and indeed the former has superseded the latter in the Shakespearean dramatic repertoire.

The fourth, and probably deepest, reason for the apparent prevalence among scientists of the proposition that artistic creations are unique and scientific creations are not can be attributed to a contradictory epistemological attitude toward the events in the outer and the inner world. The outer world, which science tries to fathom, is often viewed from the standpoint of materialism, according to which events and the relations between them have an existence independent of the human mind. Hence the outer world and its scientific laws are simply there, and it is the job of the scientist to find them. Thus going after scientific discoveries is like picking wild strawberries in a public park: the berries *A* does not find today *B* or *C* or *D* will surely find tomorrow. At the same time, many scientists view the inner world, which art tries to fathom, from the standpoint of idealism, according to which events and relations between them have no reality other than their reflection in human thought. Hence there is nothing to be found in the inner world, and artistic creations are cut simply from whole cloth. Here *B* or *C* or *D* could not possibly find tomorrow what *A* found today, because what *A* found had never been there. It is not altogether surprising, of course, to find this split epistemological attitude toward the two worlds, since of these two antithetical traditions in Western philosophical thought, materialism is obviously an unsatisfactory approach to art and idealism an unsatisfactory approach to science.

SCIENTISTS' MISCONCEPTION of the working methods of artists is idealized in this scene from the 1945 Columbia Pictures production *A Song to Remember*. Frédéric Chopin (played by Cornel Wilde), after gazing fondly at his muse George Sand (Merle Oberon), sits down at the Pleyel pianoforte and composes his "Preludes." Science, as any scientist knows, is done quite differently.

It is only in the past 20 years or so, more or less contemporaneously with the growth of molecular biology, that a resolution of the age-old epistemological conflict of materialism v. idealism was found in the form of what has come to be known as structuralism. Structuralism emerged simultaneously, independently and in different guises in several diverse fields of study, for example in psychology, linguistics, anthropology and biology.

Both materialism and idealism take it for granted that all the information gathered by our senses actually reaches our mind; materialism envisions that thanks to this information reality is mirrored in the mind, whereas idealism envisions that thanks to this information reality is constructed by the mind. Structuralism, on the other hand, has provided the insight that knowledge about the world enters the mind not as raw data but in already highly abstracted form, namely as structures. In the preconscious process of converting the primary data of our experience step by step into structures, information is necessarily lost, because the creation of structures, or the recognition of patterns, is nothing else than the selective destruction of information. Thus since the mind does not gain access to the full set of data about the world, it can neither mirror nor construct reality. Instead for the mind reality is a set of structural transforms of primary data taken from the world. This transformation process is hierarchical, in that "stronger" structures are formed from "weaker" structures through selective destruction of information. Any set of primary data becomes meaningful only after a series of such operations has so transformed.it that it has become congruent with a stronger structure preexisting in the mind. Neurophysiological studies carried out in recent years on the process of visual perception in higher mammals have not only shown directly that the brain actually operates according to the tenets of structuralism but also offer an easily understood illustration of those tenets.

Finally, we may consider the relevance of structuralist philosophy for the two problems in the history of science under discussion here. As far as prematurity of discovery is concerned, structuralism provides us with an understanding of why a discovery cannot be appreciated until it can be connected logically to contemporary canonical knowledge. In the parlance of structuralism, canonical knowledge is simply the set of preexisting "strong" structures with which primary scientific data are made congruent in the mental-abstraction process. Hence data that cannot be transformed into a structure congruent with canonical knowledge are a dead end; in the last analysis they remain meaningless. That is, they remain mean-

ARTISTS' MISCONCEPTION of the scientific act of creation is equally unrealistic. In this scene from the 1935 Warner Brothers film *The Story of Louis Pasteur* the great scientist (played by Paul Muni) has the sudden inspiration to discover the vaccine for rabies. Art, as any artist knows, is done quite differently. Both photographs are from the Museum of Modern Art Film Stills Archive.

ingless until a way has been shown to transform them into a structure that is congruent with the canon.

As far as uniqueness of discovery is concerned, structuralism leads to the recognition that every creative act in the arts and sciences is both commonplace and unique. On the one hand, it is commonplace in the sense that there is an innate, or genetically determined, correspondence in the transformational operations that different individuals perform on the same primary data. With reference to science, cognitive psychol-

ogy has taught that different individuals recognize the same "chairness" of a chair because they all make a given set of sense impressions from the outer world congruent with the same *Gestalt*, or mental structure. With reference to art, analytic psychology has taught that there is a sameness in the subconscious life of different individuals because an innate human archetype causes them to make the same structural transformations of the events of the inner world. And with reference to both art and science structural linguistics has taught

that communication between different individuals is possible only because an innate human grammar causes them to transform a given set of semantic symbols into the same syntactic structure. On the other hand, every creative act is unique in the sense that no two individuals are quite the same and hence never perform exactly the same transformational operations on a given set of primary data. Although all creative acts in both art and science are therefore both commonplace and unique, some may nonetheless be more unique than others.

THE AUTHORS

JACOB BRONOWSKI ("The Creative Process") was born in Poland, raised in Germany and trained as a mathematician in England, where in 1933 he received his Ph.D. from the University of Cambridge. He is known not only for his work in topology and statistics but also as an administrator and man of letters. Bronowski taught mathematics at University College, Hull, until 1942. He then entered the service of the British Government, which employed him to assess bomb damage and later to work in the field of operations research. In 1945 he joined the Chiefs of Staff mission to Japan—a trip which resulted in his report on atomic-bomb damage in Nagasaki and in his radio play *The Journey to Japan*. Bronowski headed the Coal Research Establishment of Britain's National Coal Board. In addition to many mathematical papers, he wrote *The Poet's Defence, William Blake: A Man without a Mask, The Common Sense of Science*, and *Science and Human Values* [see "Books"; SCIENTIFIC AMERICAN, June 1958]. He is also known for his thirteen-part PBS series, *The Ascent of Man*. Bronowski died in August, 1974.

FREDERICK G. KILGOUR ("William Harvey") served for nineteen years at Yale University as Medical Librarian and then as Associate University Librarian for Research and Development. Throughout these years he was also Lecturer in History of Science and in History of Medicine, and while there he wrote the present article. Already during World War II and the years immediately following he worked with the U.S. Government on the collection and dissemination of information, and this background combined with his other diverse interests to culminate in the establish-

ment of the OCLC in Columbus, Ohio. Originally named the Ohio College Library Center, it is now the Online Computer Library Center. As founding director, he has played a principal and formative role in the computerization of library holdings.

I. BERNARD COHEN ("Newton's Discovery of Gravity") is Victor S. Thomas Professor Emeritus of the History of Science at Harvard University. Awarded his Ph.D. by Harvard in 1947, he was the first American to be given a doctorate in this field. Formerly editor of *Isis*, the quarterly journal of the History of Science Society, he is a past president of that society and also of the International Union of the History and Philosophy of Science. In addition to his teaching at Harvard he has lectured in England, France, Italy and Japan.

MARIE BOAS HALL ("Robert Boyle") was reader in the history of science at the Imperial College of Science and Technology of the University of London. Born in Massachusetts, she obtained bachelor's and master's degrees in chemistry at Radcliffe College and a Ph.D. from Cornell University. Her first work was as a technical writer with the U.S. Army Signal Corps from 1942 to 1944; she also served as a technical writer at the Radiation Laboratories of the Massachusetts Institute of Technology from 1944 to 1946. Among the institutions where she has taught the history of science are the University of Massachusetts, Indiana University and the University of California at Los Angeles. She is the author of *Robert Boyle and 17th-Century Chemistry*, published in 1958, *Robert Boyle on Natural Philosophy*, published in 1965,

and other books and articles dealing mainly with 17th- century science. Now retired, she and her husband, A. Rupert Hall, a professor at Imperial College, edited the multivolume set of *Correspondence of Henry Oldenburg*.

DENIS I. DUVEEN ("Lavoisier") was the president of a Long Island soap company. A chemist, he was born in London in 1910, graduated from Oxford University in 1929 and did research in organic chemistry at the Collège de France. He came to the U.S. in 1948, having been during World War II technical assistant to the director of an explosives factory run by the British Ministry of Supply. A general collection of alchemical and early chemical works which he assembled is now at the University of Wisconsin. Duveen had what was the most extensive collection of Lavoisier's printed works and manuscripts, which went in 1962 to the History of Science Collections at Cornell University, and he collaborated in publishing a full bibliography of the great chemist's writings.

IAN STEWART ("Gauss") is at the Mathematics Institute of the University of Warwick. After being graduated from the University of Cambridge in 1966, he received his Ph.D. in mathematics from Warwick. In 1974 he was a Humboldt Foundation Fellow at the University of Tübingen, and in 1976 he spent six months at the University of Auckland. Currently he is at Warwick. Stewart's main field of research is infinite-dimensional Lie algebras. He writes: "My interest in the history of mathematics stems from a desire to understand how the subject of mathematics fits together. Many of the cross-connections become clear only if the history is borne in mind." He has published numerous books on mathematics (including several for children).

TONY ROTHMAN ("The Short Life of Évariste Galois") wrote this article while he was a visiting fellow in astrophysics at the University of Oxford. He received his undergraduate education at Swarthmore College, from which he was graduated in 1975. In 1981 he got his Ph.D. in physics from the Center for Relativity at the University of Texas at Austin. Since then he has done postdoctoral work at the Sternberg Astronomical Institute at Moscow, the University of Capetown, as well as Oxford. His research interests include black holes, the formation of baryons early in the history of the universe and the synthesis of atomic nuclei in the Big Bang. He is also a professional writer, having published a novel and a number of popular articles on scientific subjects. He is at work on a second novel. Rothman writes: "My interest in Galois actually stemmed from a play ... I wrote a few years ago. It concerns the Russian poet Pushkin as well as Galois, In the course of my background research I discovered that the standard accounts of Galois's life in English were, to say the least, inaccurate." The play won the 1981–1982 Oxford Experimental Theatre Club Competition.

MITCHELL WILSON ("Joseph Henry") is a novelist, physicist and one-time industrial researcher who has made a name for himself as one of the few literary "regional" writers in the field of science and technology. While at New York University and Columbia University he felt an equal pull toward literature and science. Wilson did graduate work on the meson, and in 1940 joined the research staff of the Columbian Carbon Company. All the while he was also striving to be a writer, selling his first story to Cos-

mopolitan in 1939. In 1944 he found he had to make a choice between research and writing. The first product of his commitment to the latter was *Live with Lightning,* a novel which got some critical acclaim as the story of how it is to be a physicist in these times. Since then Wilson has written two novels. Wilson feels that technology and its men now form the background and hard core of American living, as authentic as the Western plains and mountains of an earlier tradition. He has worked on a continuation of his inventors' story, as well as on a history of American science and invention.

LOREN EISELEY ("Charles Darwin") is professor of anthropology at the University of Pennsylvania. He was a frequent contributor to SCIENTIFIC AMERICAN. A redoubtable and a passionate essayist, Eiseley's works included *The Immense Journey* (1957); *The Unexpected Universe* (1969); *The Invisible Pyramid* (1970); and *The Night Country* (1971). In 1975 he published an autobiography entitled *All the Strange Hours*. He worked on a project, which entailed assembling material on evolution to shed light on the correspondence between Darwin and Sir Charles Lyell, which the American Philosophical Society purchased. Eiseley also wrote a history of evolutionary thought for the Doubleday Anchor Book series. He died in July, 1977.

A. HALLAM ("Alfred Wegener and the Hypothesis of Continental Drift") teaches geology and mineralogy at the University of Oxford. His bachelor's degree and his Ph.D. are from the University of Cambridge. He has taught at the University of Edinburgh, Stanford University and McMaster University in Ontario. The author of an earlier article in SCIENTIFIC AMERICAN ("Continental Drift and the Fossil Record," November, 1972), Hallam writes that his current interests "are in the field of Jurassic geol-

ogy and palaeontology, with my interest in the fossils being (a) patterns of evolution, (b) ecology and (c) biogeography."

DANIEL J. KEVLES ("Robert A. Millikan") is professor of history at the California Institute of Technology. He studied physics at Princeton University and went on to obtain his Ph.D. in history there in 1964. After working for a summer on the White House staff he joined the Cal Tech faculty. His historical research has focused on American science in the 19th and 20th centuries, and he has written a history of genetics and eugenics in the U.S. and Britain.

GUNTHER S. STENT ("Prematurity and Uniqueness in Scientific Discovery") is professor of molecular biology and of bacteriology at the University of California at Berkeley. He also wrote "Cellular Communication" in the SCIENTIFIC AMERICAN September, 1972 single-topic issue on communication. Concerning the genesis of the present article, he writes: "In May of 1970 the American Academy of Arts and Sciences held a small conference in Boston on the History of Biochemistry and Molecular Biology at which I was asked to make a few brief comments following an account by Salvador Luria of the origins of molecular genetics. I intended to speak for about five minutes, a time I thought was more than enough to point out the relevance of the prematurity and the uniqueness concepts to Luria's reminiscences about Oswald Avery and James Watson. But the conference participants—both scientists and historians—kept on interrupting me, and my 'brief comments' eventually lasted twice as long as Luria's lecture. This article is the product of that vigorous discussion, and among the discussants, I am especially indebted to Robert K. Merton and Harriet A. Zuckerman of Columbia University for helping me to focus my ideas more sharply."

BIBLIOGRAPHIES

1. The Creative Process

THE ORIGINS OF MODERN SCIENCE, 1300–1800. H. Butterfield. G. Bell and Sons Ltd., 1949.

THE PHYSICAL WORLD OF THE GREEKS. S. Sambursky. Routledge and Kegan Paul Ltd., 1956.

SCIENCE AND HUMAN VALUES. J. Bronowski. Julian Messner, Inc., 1958.

2. William Harvey

EXERCITATIO ANATOMICA DE MOTU CORDIS ET SANGUINIS IN ANIMALIBUS. William Harvey. With an English translation and annotations by C. D. Leake, Charles C Thomas, 1928.

THE LIFE OF WILLIAM HARVEY. Sir Geoffrey Keynes. Oxford University Press, 1966.

WILLIAM HARVEY, THE MAN, THE PHYSICIAN AND THE SCIENTIST. Kenneth D. Keele. London, 1965.

3. Newton's Discovery of Gravity

THE BACKGROUND TO NEWTON'S PRINCIPIA: A STUDY OF NEWTON'S DYNAMICAL RESEARCHES IN THE YEARS 1664–84. John W. Herivel. Oxford University Press, 1965.

FROM KEPLER'S LAWS, SO-CALLED, TO UNIVERSAL GRAVITATION: EMPIRICAL FACTORS. Curtis A. Wilson in Archive for History of Exact Sciences. Vol 6, No. 2, pages 89–170; April 29, 1970.

FORCE IN NEWTON'S PHYSICS: THE SCIENCE OF DYNAMICS IN THE SEVENTEENTH CENTURY. Richard S. Westfall. American Elsevier Publishers Inc., 1971.

THE NEWTONIAN REVOLUTION: WITH ILLUSTRATIONS OF THE TRANSFORMATION OF SCIENTIFIC IDEAS. I. Bernard Cohen. Cambridge University Press, 1980.

4. Robert Boyle

FROM GALILEO TO NEWTON, 1630–1720. Rupert Hall. Harper & Row, 1963.

ROBERT BOYLE AND SEVENTEENTH CENTURY CHEMISTRY. Marie Boas. Cambridge University Press, 1958.

ROBERT BOYLE ON NATURAL PHILOSOPHY. Marie Boas Hall. Indiana University Press, 1965.

5. Lavoisier

ANTOINE LAVOISIER: SCIENTIST, ECONOMIST, SOCIAL REFORMER. Douglas McKie. Henry Schuman, Inc., 1952.

A BIBLIOGRAPHY OF THE WORKS OF ANTOINE LAURENT LAVOISIER, 1743–1794. Denis I. Duveen and Herbert S. Klickstein. Wm. Dawson & Sons, Ltd., 1954.

LAVOISIER. Henry Guerlac in Dictionary of Scientific Biography, edited by Charles Coulston Gillispie. Charles Scribner's Sons, 1972.

6. Gauss

MEN OF MATHEMATICS. Eric T. Bell. Simon & Schuster, Inc., 1937.

DISQUISITIONES ARITHMETICAE. Carl Friedrich Gauss, translated by Arthur A. Clarke. Yale University Press, 1966.

CARL FRIEDRICH GAUSS: A BIOGRAPHY. Tord Hall. The MIT Press, 1970.

WERKE. Carl Friedrich Gauss. G. Olms Verlag. Hildesheim–New York, 1973.

7. The Short Life of Evariste Galois

LA VIE D'ÉVARISTE GALOIS. Paul Dupuy in Annales scientifiques de l'École normale supérieure, Vol. 13, pages 197–266; 1896.

ÉCRITS ET MÉMOIRES MATHÉMATIQUES D'ÉVARISTE GALOIS. Edited by Robert Bourgne and J.-P. Azra. Gauthier-Villars, 1962.

ÉVARISTE GALOIS. René Taton in Dictionary of Scientific Biography, edited by Charles Coulston Gillispie. Charles Scribner's Sons, 1972.

GENIUS AND BIOGRAPHERS: THE FICTIONALIZATION OF ÉVARISTE GALOIS. Tony Rothman in The American Mathematical Monthly, Vol. 89, No. 2, pages 84–106; February, 1982.

René Taton has sent me an extract from André Dalmas's biography of Galois. (André Dalmas, Évariste Galois, Révolutionnaire et Géomètre, [Paris: Fasquelle, 1956].) It contains an article which appeared in the June 4, 1832, issue of the Lyon "journal constitutionnel" Le Précurseur. If accurate, this account is as close as we are ever likely to get to the truth about Galois's death. As published by Dalmas the report reads:

Paris, 1 June — A deplorable duel yesterday has deprived the exact sciences of a young man who gave the highest expectations, but whose celebrated precocity was lately overshadowed by his political activities. The young Évariste Galois, condemned for a year as a result of a toast proposed at a banquet at the Vendanges de Bourgogne, was fighting with one of his old friends, a young man like himself, like himself a member of the Society of Friends of the People, and who was known to have figured equally in a political trial. It is said that love was the cause of the combat. The pistol was the chosen weapon of the adversaries, but because of their old friendship they could not bear to look at one another and left the decision to blind fate. At point-blank range they were each armed with a pistol and fired. Only one pistol was charged. Galois was pierced through and through by a ball from his opponent; he was taken to the Cochin Hospital where he died in about two hours. His age was 22. L. D., his adversary, is a bit younger.

Dalmas goes on to write:

> Discarding the error about Galois's age, the report of the journalist appears plausible. There was only one young republican who figured with Galois in a political trial: Duchatelet. This is confirmed moreover by the initial D. These new particulars render the hypothesis of a [police or political] provocation very uncertain indeed. (Ibid., pp. 77–78. My translation and brackets.)

I have several comments. The report was mistaken on both the date of the duel, which took place on May 30, and the date of Galois's death, which was May 31 at about 10 o'clock in the morning. Duchatelet was Galois's republican friend. They were apprehended together for wearing uniforms of the Artillery of the National Guard on Bastille Day 1831. An obvious point to check is whether Duchatlet's first initial was L., but according to Bourgne and Azra (Robert Bourgne and J. P. Azra, eds., *Écrits et Mémoires Mathématiques d'Évariste Galois: Édition Critique Intégral de ses Manuscrits et Publications*, [Paris: Gauthier-Villars, 1962], p. XXVIII.) Duchatelet's given name was Ernest. In addition, there is Dumas's statement that Galois was killed by Pescheux d'Herbinville whose surname, bearing in mind the variable orthographic procedures of the times, might also be said to begin with a D (e.g., Dumotel versus du Motel). D' Hebinville also figured in a political trial but not with Galois.

To what conclusions does this new information lead? Galois wrote in one of his last letters, "I am the victim of an infamous coquette and her two dupes." (Ibid., p. 470.) The first possibility is that both the newspaper article and Dumas

are substantially correct and we have now identified both of the "dupes." The other possibilities are that either Dumas or the *Précurseur* journalist or both are mistaken. Undoubtedly, *Le Précurseur* should be accorded more weight—Duchatelet is the best candidate. But I do not think the final identification of duelist is terribly important, for *all* contemporary reports agree that the duel was between friends, for nonpolitical reasons, and that there was no police intervention of any kind.

8. Joseph Henry

JOSEPH HENRY, HIS LIFE AND WORK. Thomas Coulson. Princeton University Press, 1950.

THE EXPLORING EXPEDITION AND THE SMITHSONIAN INSTITUTION. Nathan Reingold and Marc Rothenberg, in *Magnificent Voyagers: The U. S. Exploring Expedition*, edited by Herman J. Viola and Carolyn Margolis. Smithsonian Institution Press, 1985.

AMERICA'S CASTLE. Kenneth Hafertape. Smithsonian Institution Press, 1985.

HENRY. Nathan Reingold in *Dictionary of Scientific Biography*, edited by Charles Coulston Gillispie. Charles Scribner's Sons, 1972.

9. Charles Darwin

THE FOUNDATIONS OF THE ORIGIN OF SPECIES. Charles Darwin. Cambridge University Press, 1909.

THE DARWINIAN HERITAGE. Edited by David Kohn. Princeton University Press, 1986.

THE BEAGLE RECORD. Edited by Richard Darwin Keyes. Cambridge University Press, 1979.

AUTOBIOGRAPHY OF CHARLES DARWIN. With notes by Nora Barlow. 1958, reprint Norton, 1969.

DARWIN. Peter Brent. Harper and Row, 1981.

CHARLES DARWIN, A PORTRAIT. Geoffrey West (pseudonym for Geoffrey Wells). Yales University Press, 1938.

A CALENDAR OF THE CORRESPONDENCE OF CHARLES DARWIN. Edited by Frederick Burkhardt and Sydney Smith. Garland, 1985.

APES, ANGELS, AND VICTORIANS: THE STORY OF DARWIN, HUXLEY, AND EVOLUTION. William Irvine. 1955, reprint 1983, University Press of America.

10. Alfred Wegener and the Hypothesis of Continental Drift

ALFRED WEGENER. Else Wegener. F. A. Brockhaus, Wiesbaden, 1960.

MEMORIES OF ALFRED WEGENER. J. Georgi in *Continental Drift*, edited by S. K. Runcorn. Academic Press, 1962.

CONTINENTAL DRIFT: THE EVOLUTION OF A CONCEPT. Ursula B. Marvin. Smithsonian Institution Press, 1973.

A REVOLUTION IN THE EARTH SCIENCES: FROM CONTINENTAL DRIFT TO PLATE TECTONICS. A. Hallam. Oxford University Press, 1973.

11. Robert A. Millikan

THE AUTOBIOGRAPHY OF ROBERT A. MILLIKAN. Robert A. Millikan. Prentice-Hall, Inc., 1950.

COSMIC RAYS. Bruno Rossi. McGraw-Hill Book Company, 1964.

THE ELECTRON: ITS ISOLATION AND MEASUREMENT AND THE DETERMINATION OF SOME OF ITS PROPERTIES. Robert A. Millikan. The University of Chicago Press, 1968.

THE CONSERVATIVE MODE: ROBERT A. MILLIKAN AND THE TWENTIETH-CENTURY REVOLUTION IN PHYSICS. Robert H. Kargon in *Isis*, Vol. 68, No. 244, pages 509–526; December, 1977.

THE PHYSICISTS. Daniel J. Kevles. Alfred A. Knopf, Inc., 1978.

12. Prematurity and Uniqueness in Scientific Discovery

THE POTENTIAL THEORY OF ADSORPTION. Michael Polanyi in *Science*, Vol. 141, No. 3585, pages 1010–1013; September 13, 1963.

PERCEPTION AND DECEPTION. C. W. Churchman in *Science*, Vol. 153, No. 3740, pages 1088–1090; September 2, 1966.

MOLECULAR GENETICS: AN INTRODUCTORY NARRATIVE. Gunther S. Stent. W. H. Freeman and Company, 1971.

INDEX